CW00552385

I
Am a
Mathematician

I
Am a
Mathematician

The Later Life of a Prodigy

An autobiographical account of
the mature years and career of

Norbert Wiener

and a continuation of the account
of his childhood in *Ex-Prodigy*.

 The M.I.T. Press
Massachusetts Institute of Technology
Cambridge, Massachusetts, and London, England

Sixth Printing, 1989
First MIT Press paperback edition, August 1964

ISBN 0 262 73007 3 (paperback)
Library of Congress Catalog Card No: 56-5598
Printed in the United States of America

To the Massachusetts Institute of Technology, which has given me the encouragement to work and the freedom to think.

Contents

1   My Start as a Mathematician   17

2   The International Mathematical Congress of 1920 at Strasbourg   44

3   1920–1925. Years of Consolidation   71

4   The Period of My Travels Abroad—Max Born and Quantum Theory   89

5   To Europe as a Guggenheim Fellow with My Bride   110

6   1927–1931. Years of Growth and Progress   124

7   An Unofficial Cambridge Don   147

8   Back Home. 1932–1933   165

Contents

9   Voices Prophesying War. 1933–1935   173

10   China and Around the World   183

11   The Days before the War. 1936–1939   208

12   The War Years. 1940–1945   226

13   Mexico. 1944   276

14   Moral Problems of a Scientist. The Atomic Bomb. 1942–   293

15   Nancy, Cybernetics, Paris, and After: 1946–1952   314

16   India. 1953   338

17   Epilogue   357

Index   367

Preface

I have just finished dictating the last word of this book. It represents that part of my autobiography which dates roughly from my arrival at M.I.T. in 1919, when I was twenty-four years old. The earlier part of my autobiography, under the title *Ex-Prodigy*, concerned my childhood and adolescence, while the present book is occupied with my mature personal and scientific career.

It deals with my work, my travel, and my personal experiences, and I hope that it will give a fair account of my intellectual development. My problem has been to explain to a public that is generally not too deeply interested in science, and which is certainly not possessed of a technical acquaintance with it, the development of ideas which are fundamentally scientific. I have had as far as possible to eschew the technical vocabulary of science and to translate my concepts into everyday language. This is a splendid discipline for an author, but it also is a discipline which runs the risk of falling short of full success. While the use of scientific terms often results in jargon, to express any significant part of scientific ideas without the compact denotation and connotation which the history of science has given these terms is most difficult and is much less likely to be completely successful than the specifically literary critic may be aware.

Thus, I have had two years discipline at a piece of work in which, by the nature of things, I must wait for the judgment of the public in order to be sure that I have achieved a measure of success. Why did I assume this uninviting labor, which at the best can add little to my stature as a working scientist and at the worst must offer new opportunities for those who may be inclined to criticize me? All in all, I don't know. There certainly have been motives of literary vanity and the desire to show that, both as an individual and as a scientist I have been able to accomplish a task off my regular beat.

Admitting this freely, there are other, more important motives. As in the first volume of my autobiography, so here, too, I wish to think out to myself what my career has meant and to come to that emotional peace which only a thorough consideration and understanding of one's own past can bring. I have also wished to make this understanding available to young men coming up through similar careers in mathematics and the other sciences. I have felt that the scientist, his mode of life, and the demands on him are not sufficiently known to the larger intellectual public, and I believe that here I have the duty of exegesis. Then, finally, I have not had any previous opportunity to write up many ideas—treated singly in my existing literary work—in the form of a consistent historical account of how I came by them.

I should like to mention, among the colleagues with whom I

have discussed this undertaking, Professor Karl Deutsch of the M.I.T. Department of Humanities; Professor Armand Siegel of Boston University; Dr. Dana L. Farnsworth, formerly of the Medical Department at M.I.T. and now professor of hygiene at Harvard University; and Dr. Morris Chafetz. In addition, I wish to give my thanks to the several members of the sequence of secretaries who have taken my dictation, who have expressed their criticism of my ideas, and who have helped put my work into printable form.

Norbert Wiener

Cambridge, Massachusetts
Washington's Birthday, 1955

I
Am a
Mathematician

1

My Start

as

a Mathematician

This book is the second volume of my autobiography. The first volume, entitled *Ex-Prodigy,* was devoted to my early education, to my relations with my father, and to the unusual experience of being an infant prodigy. The present book is devoted to my career as a working mathematician.

For purposes of organization, I must begin this volume at some definite period, and the natural time to begin it is in 1919, when I first came to the Massachusetts Institute of Technology. I was then twenty-four years old: too old to be an infant prodigy any longer but not too old to have the marks of having been one firmly impressed upon me.

I do not intend to go back to my childhood, which was the subject of my earlier book, to explain the origins of these marks, but I must report certain features of my adolescence which in the previous volume represented the last stages in the life of a child but here are rather to be interpreted as the first stages in the life of a man.

Of all the influences which operated on me during my childhood and adolescence, the most important was my father, Leo Wiener, professor of Slavic languages and literatures at Harvard University—a small, vigorous man, of emotions both deep and quick, sudden in his movements and his gestures, ready to approve and to condemn, a scholar rather by nature than by any specific training. In him were joined the best traditions of German thought, Jewish intellect, and American spirit. He was given to overriding the wills of those about him by the sheer intensity of his emotion rather than by any particular desire to master other people.

Having been more than twenty years in the shadow of such a man, with the knowledge that I was flesh of his flesh and bone of his bone, I myself was cast into a mold sharply different from that of most of those about me. I became a scholar partly because it was my father's will but equally because it was my internal destiny. From my earliest years I had been interested in the world about me and very inquisitive into its nature. I had learned to read by the time I was four, and almost from then on I plunged into scientific reading of the most varied character. By the time I was seven, my reading had come to range from Darwin and Kingsley's Natural History to the psychiatric writings of Charcot and Janet and others of the school of the Salpêtrière. This miscellaneous collection of learning had been assembled in those remarkable and diversified volumes of small and blunt print known as the Humboldt Library.

My own free curiosity was matched by my father's insistence that my training be disciplined. I learned my science for myself; my father introduced me to languages, both ancient and modern, and to mathematics. All these subjects had a certain interest for me, but no casual interest could satisfy my father's demands for precise and ready knowledge. These demands were severe and painful, but they were made tolerable by my complete awareness of my father's integrity and intellectual power.

The arduous course of training to which I was put tended to isolate me from the world and to give me a certain aggressive, unlovable naïveté. I played a good deal with other boys, but I was not greatly welcomed by them. When I entered high school, at the age of nine, the few companions that I found were not among the high-school boys but among children of my own age.

The special nature of my position was aggravated by a near-sightedness which at one time seemed to threaten my vision altogether. While this had no direct effect on my physical vigor, it cut me out of that whole sector of boyish life which depends on skill at games. It also tended to accentuate my very marked physical clumsiness. This clumsiness was serious enough on its own merits, but it was further brought out by the way in which my father harped on it and used to humiliate me concerning it. He himself was no prodigy of manual skill, but he was interested in farming, gardening, and an outdoor life, and he used his limited skills to the best possible advantage. His discomfiture at my shortcomings was real.

With the inevitable isolation which my father's training gave me, I was a very self-conscious hobbledehoy, subject to alternate moods of conceit when I became aware of my abilities and of great disappointment when I accepted at their face value my father's strictures on my shortcomings, or when I contemplated the long and uncertain road to achievement to which my highly eccentric bringing-up had condemned me. Moreover, I had my father perpetually before me as an example of a certain admissible aggressiveness, although its naturalness and to some extent its justifiability made it less devastating to me than it would otherwise have been.

Superimposed upon this was another source of unsureness

which continued to haunt me for a large part of my life. My mother, like my father, was a Jew. Unlike my father, she resented being a Jew. My father and mother concurred, it is true, in assimilationism and in a desire to identify themselves and their children with the general American environment. But while this desire took a defensible form with my father, whose many interests allowed him to view this fact of our origin in something like a proper perspective, anti-Semitism became, in all its starkness, the chief subject of my mother's concern.

On the one hand, we children were brought up not only in ignorance of our origins but under a directly false impression of them. On the other hand, we could at no time fully fail to realize that there was some unexplained Jewish element in our environment. My mother made many uncomplimentary remarks concerning the Jews, which went far to impress on me that she considered her Jewish origin, and consequently our own Jewishness, a source of shame. When, later—at the age of fifteen—I learned from my father that we were unquestionably Jewish, remembered remarks of my mother forced on me a sense of inferiority which contributed greatly to my insecurity so that it was a matter of many years before I could acquire a decent measure of self-confidence. Thus, I was subject to an alternation of awareness of my powers and doubts as to my value; and these made me continue to oscillate between an unlovely conceit and an equally unlovely abjectness.

There were, however, certain important factors in my make-up which made for success in general and for intellectual success in particular. My father's independence had been reflected in both my nature and my habits. His power did not consist merely in a high level of intellectual ability but in a willingness to supplement this ability by hard and unceasing work. I had seen my father bring himself to a breakdown by the herculean task of translating twenty-four volumes of Tolstoy in two years. What Father expected of himself he expected of me as well, and from childhood on I have known no period when I was content to rest on past accomplishments.

From high school I proceeded to Tufts College, near Boston, and later to graduate work at Harvard and at Cornell. I took my bachelor's degree at Tufts at the age of fourteen and my doctor's degree in philosophy at Harvard at eighteen. As I gradually acquired a limited amount—a very limited amount— of independence from my father, I found that the dawning freedom of approaching manhood was largely a freedom to make mistakes and to know failure. Yet even this joyful freedom was limited by my father's proneness to make sudden decisions affecting my whole future which bound me as much as if these decisions had been my own.

After obtaining my Ph.D. I was appointed by Harvard University to a traveling fellowship which I spent at Cambridge, England, and later at Göttingen, in Germany. Although I had been away from home before, this was the first time that I had come to be really competent to live alone and could learn something about the freedom of an independent worker. My chief teacher and mentor was Bertrand Russell, with whom I studied mathematical logic and a good many more general matters concerning the philosophy of science and mathematics. Russell, who looked then, as he does now, like the Mad Hatter, gave beautifully finished discussions largely devoted to Einstein's new theory of relativity. I studied Russell's own mathematical-logic writings with a small group of students who met in his rooms, and I followed other courses which he recommended to me. The chief of these was a higher-mathematics course under G. H. Hardy, later to become a professor both at Oxford and at Cambridge and perhaps the greatest figure of his mathematical generation in England.

My doctor's thesis at Harvard had been on the philosophy of mathematics. Russell impressed upon me that to do compe-

tent work in the philosophy of mathematics I should know
more than I did about mathematics itself.

Hardy, to whom I turned, was an ideal mentor and model for
an ambitious young mathematician. I had first seen him in
Russell's rooms when my father took me to Cambridge and left
me in an absent-minded way to sink or swim there. At that
time, Father and I both took Hardy to be an undergraduate,
a shy, self-effacing young man who, as I came to learn later,
was an excellent athlete and an authority on every game played
with a ball. In his later years he was to become a rather dried-
up, wizened figure, in the inevitable unpressed jacket and bags
of a Cambridge don: kindly and helpful but detached and
desperately afraid of women, and it is in this way that I remember him best.

His course was a delight to me. My previous adventures into
higher mathematics had not been completely satisfying, because I sensed gaps in many of the proofs which I was unwilling to disregard—and correctly too, as it later turned out,
for the gaps were really there and they should have disturbed
not only me but my earlier teachers. Hardy, however, led me
through the complicated logic of higher mathematics with such
clarity and in such detail that he resolved these difficulties as
we came to them and gave me a real sense of what is necessary
for a mathematical proof. He also introduced me to the Lebesgue integral, which was to lead directly to the main achievements of my early career.

The Lebesgue integral is not an easy conception for the
layman to grasp, but since an awareness of it is fundamental to this book, I shall try to suggest, if not its full complexity,
at least its main theme. It is easy enough to measure the length
of an interval along a line or the area inside a circle or other
smooth, closed curve. Yet when one tries to measure sets of
points which are scattered over an infinity of segments or
curve-bound areas, or sets of points so irregularly distributed

that even this complicated description is not adequate for them, the very simplest notions of area and volume demand high-grade thinking for their definition. The Lebesgue integral is a tool for measuring such complex phenomena.

The measurement of highly irregular regions is indispensable to the theories of probability and statistics; and these two closely related theories seemed to me, even in those remote days before the war, to be on the point of taking over large areas of physics. They stood approximately in the middle ground where physics and mathematics meet, and this middle ground was just where I was eventually to do my best work, for such work seemed to be in harmony with a basic aspect of my own personality.

Even more, they anticipated the main tendencies of modern mathematics, grounded in physics and dependent on the ideas of measure and probability expounded in the statistical mechanics of that great American scientist, Josiah Willard Gibbs. The development of problems surrounding the application of mathematical ideas into problems that arise in the physical world constitutes one of the main themes of this volume.

The theory of the Lebesgue integral leads the student from the measure of intervals to the measure of more complex phenomena obtained by combining sequences of intervals, and then to sets which can be approached by such sequences, while the sets of points excluded from them can be approached in a similar manner. There is nothing in it which can be explained satisfactorily to the layman, but neither is there anything which involves an excessive complication of logic and manipulation. It enabled Lebesgue to extend the notion of length or measure from the single interval to the extreme significant limits at which measure is possible.

Hardy died some years ago, but his younger colleague and research partner, J. E. Littlewood, with whom I also worked, is still alive. At that time, Littlewood seemed to me merely a bright young man among other bright young men, but, as I got to know him later, I learned that he was a great rock climber as well as a mathematician. He had the small, muscular, compact build of the true rock climber, and both as a rock climber and a mathematician he showed an unlimited power and an impeccable technique.

In their long partnership, Hardy and Littlewood occupied easily distinguishable roles, for whereas Hardy was the man of clarity and original ideas, Littlewood was the man of power and indomitable persistence. Curiously enough, Littlewood was the more self-effacing of the two. Later on, when he visited Edmund Landau at Göttingen, that irrepressible, spoiled child of mathematics said to him, "So you do exist! I thought you were merely a name used by Hardy for those papers which he didn't think were quite good enough to publish under his own name."

Landau and David Hilbert became my teachers later the same academic year, when I studied in Göttingen in the spring term of 1914, just as the First World War was about to break out. Landau came from a rich Jewish banking family and had been something of a prodigy himself. He had been brought up with every luxury which his wealthy parents were able to provide. A small, cherubic figure with a bristling little mustache and a completely undisciplined conceit, he always seemed a bit out of place in the real world. When people asked where to find his house in Göttingen he would say quite naïvely, "You will have no difficulty in finding it. It is the finest house in the town."

Hilbert, with whom I also studied, was a very different sort of person, a quiet, peasantlike man from East Prussia. He was conscious of his strength but genuinely modest. He used to say about his son, who certainly lacked the powers of his father, "He has his mathematical ability from his mother, everything else from me."

Hilbert himself had taken up, in succession, most of the difficult problems in every field of modern mathematics, and in each field he had made a major killing. He represented the grand tradition in the mathematics of the beginning of the century. For me as a young man he became the sort of mathematician whom I would have liked to become, combining tremendous abstract power with a down-to-earth sense of physical reality.

At Cambridge, Russell had impressed on me not only the importance of mathematics but the need for a physical sense, and he had suggested that I study the new developments of Rutherford and others concerning the theory of the electron and the nature of matter. At that time I did not make much headway in these subjects, but at least I was given a preview of that theory of the atomic nucleus which has since given rise to the transmutation of the elements and the construction of the atomic bomb. This theory has come to occupy a position alongside Einstein's theory of relativity, the importance of which I also learned from Russell. I found Russell's respect for physics reinforced in the scientific atmosphere of Göttingen.

After Göttingen I came back to New Hampshire for the summer just as World War I was breaking out. I returned to England for the next academic year at Cambridge, but in that atmosphere of calamity and doom I did not find many people with a heart to do serious scientific thinking nor was I myself able to carry on to any very good effect. In the late winter of 1914–1915 the German submarine campaign became threatening, and I was summoned home by my father.

The war took some years to come to America, but it was never out of my thoughts. The present generation, which has been brought up with crisis as its daily associate, can scarcely be aware of the shock with which the war impinged on my contemporaries. We had been brought up to consider the long Victorian peace as normal, and we had hoped for a continued and slow evolution towards better conditions. Not even now, forty years after, have we really been able to accept the prolonged succession of catastrophes about us as normal. I am afraid that we all have from time to time a sneaking hope that we shall wake and find ourselves again in the dull, mild days of the beginning of the century.

During the early period of the First World War I carried on a number of activities, scientific and personal, in a desultory way. I had a certain idea at the back of my mind that the war would be over some day fairly soon and that then we could return to our old settled habits of living and make long plans. I finished the incomplete academic year of 1914–15 at Columbia, listening impatiently and impolitely to various professors who did not impress me after my experience with Hardy and the Harvard philosophers. I was a pest to the older men in the graduate dormitory with my self-assertiveness and bad bridge-playing, and I filled in my abundant but lonely leisure with long walks from Columbia University to the Battery, with an overdose of films and theater in between.

My mathematical work was active but abortive. I made some attempts to apply the abstract way of thought which I had learned from Russell to topology, that strange branch of mathematics dealing with knots and other geometric shapes whose fundamental relations are not changed even by a thorough kneading of space so long as nothing is cut and no two remote points of space are joined. Topology includes the study of such things as the familiar one-sided Möbius's sheet of paper, which you get when you take a long, flat strip, rotate one end of it through half a revolution, and glue the ends. It makes an excellent parlor trick to ask a layman what will happen to such a strip if you start cutting it down the middle until the ends of the cut meet. If you try this, you will find that even after the cut is complete the strip will remain in one piece but will now make a full revolution instead of half a revolution as you proceed around it.

Not many years later topology became *the* fashionable branch of mathematics, particularly in America, under the guidance of Oswald Veblen and J. W. Alexander. But by this time I had grown disappointed in the slowness in which my work led to positive results and had destroyed or lost the manuscript on topology which I had worked on at Columbia.

In the academic year 1915–16 I returned to Harvard as docent lecturer and assistant, and I gave a series of lectures in accordance with the peculiar clause in the University statutes which conveys to every Ph.D. the privilege of giving such lectures strictly on his own responsibility. The subject I chose had to do with the work of Alfred North Whitehead, and my lectures were devoted to showing how mathematics might be based on processes of logical construction. Whitehead had shown by examples how such methods might secure for various mathematical concepts those properties which another mathematical school, that of the postulationalists, had sought as the consequences of rather arbitrary formal assumptions. For example, it was Whitehead who had thought of representing a point as the set of all areas which, according to more usual mathematical language, would be said to contain this point. But my lectures ran into certain logical difficulties which were clearly pointed out to me by Professor G. D. Birkhoff of Harvard, to whom I shall have many occasions to refer later in this book.

He was a lean, tall, Michigan Dutchman, with the drawn face and tight mouth of a rigid Calvinist, and the first important American mathematician to have had all his training in America. He had written a brilliant dissertation on certain branches of dynamics, concerning in particular the mechanics of the planets, a field which had been outlined in France by

Henri Poincaré. Birkhoff was fully conscious of his really great powers and was determined to become and to remain the first American mathematician in those classical branches of mathematics known as analysis, which constitute the extension and the elaboration of Newton's calculus and physics.

He was, as I was later to learn, intolerant of possible rivals, and even more intolerant of possible Jewish rivals. He considered that the supposed early maturity of the Jews gave them an unfair advantage at the stage at which young mathematicians were looking for jobs, and he further considered that this advantage was particularly unfair, as he believed that the Jews lacked staying power. At the beginning I was too unimportant a youngster to attract much of his attention, but later on, as I developed more strength and achievement, I became his special antipathy, both as a Jew and, ultimately, as a possible rival.

At the time at which I had first felt Birkhoff's hostility I had not been fully aware of all the forces contributing to it and lying within me and my immediate environment. I have suggested that I was not a very amiable young man. Frankly it was scarcely to have been expected that I should have been amiable. I have spoken at length about the fact of my having been a prodigy, and I need not say much about the distaste which established people felt when they were confronted with a youngster whom they did not know how to place. A career devoted to scientific achievement from the years of childhood leaves very little time for the cultivation of the social graces.

Even with all these things taken into account, I was an aggressive youngster. I felt at the bottom of my consciousness that it would take all the aggressiveness I might have to force success from the very ambiguous situation in which I found myself. Moreover, my father—who, notwithstanding all the elements of conflict between us, was my ideal and my closest mentor—was himself a very aggressive personality.

I had heard many echoes of the somewhat trivial disagreements which my father had had with his colleagues, but I was not fully aware of the fact that in addition to these individual instances he had been regarded as an essentially quarrelsome man. Part of this reputation of his was justified, but an even more important part was due to a misunderstanding of his mercurial temperament by less mobile personalities. Many years after the event, I have come to learn that a not inappreciable element in Birkhoff's antagonism toward me was due to his misunderstanding of my father and to the distaste with which he received my father's somewhat uninhibited boastings concerning myself.

The next year found me at the University of Maine, where I had secured a job through a teachers' agency. I felt humiliated that I could not get a job directly on the prestige of my academic record. The tribulations of this period of what I considered an exile belong to my previous volume and have been discussed there.

At the end of the academic year the United States entered the war. I tried unsuccessfully to join one of the services but was rejected everywhere because of my poor eyesight. I worked for a brief period at the General Electric works at Lynn, from which job I was withdrawn by my father for what he considered the better offer which I received: that of a hack writer on the Encyclopedia Americana at Albany.

I left this job to work, together with a mixed bag of civilian and military mathematicians, at the Aberdeen Proving Ground in Maryland. The work concerned problems of the design of range tables for artillery weapons. Here I spent more than half a year, first as a civilian and then as a soldier, where I fared rather poorly since the fact of having been an infant prodigy led me to commit many blunders which were not intended in a vicious spirit but which may have left a disagreeable impression of my whole personality. I was desperately unhappy under barracks conditions and did nothing to endear myself to my companions. I was honorably discharged from the service in February 1919.

After several months of newspaper hack writing, I composed a couple of scientific articles on algebra which were good enough in their own way, but which have remained completely off the beaten track. Then Professor W. F. Osgood of Harvard secured me an appointment as instructor in the department of mathematics of the Massachusetts Institute of Technology.

Osgood was a friend of my father's, and his sons had been to some slight extent my playmates years before. He was perhaps the chief representative of the German tradition in American mathematics, having studied at Göttingen, where he married a German girl, and brought back the determination to live in America the life of a German professor. Perhaps I should say the life of a German privy councilor, because his model in all things was Felix Klein, who had been for years the pope of German mathematics and had enjoyed the exalted title of Geheimrat. Osgood was a sturdy man of ruddy complexion, already getting very bald, with a spreading spade beard after the European manner. He used to pontificate at the meetings of the Harvard Mathematics Club, where he smoked his cigar after a fashion which he had obviously learned from someone else, and which we later found he had got from Felix Klein himself. He would spear it with his penknife and smoke the stub to the bitter end.

He wrote some of his books in a German of a very tolerable quality and had strong quasi-moralistic ideas of what was right and what was wrong in mathematics quite apart from any question of simple logical correctness. Those who worked under him found that he considered them bound by these ideas.

I have perhaps been insufficiently grateful to Professor Osgood for the really good turn he did me in securing me the call to the Massachusetts Institute of Technology—or M.I.T., as it is more frequently known. There were, however, certain offsets to this act of kindness. For one thing, I never felt that I had earned any real esteem from him, nor did I feel that he had made me welcome at Harvard. Furthermore, jobs were plentiful with the resumption of normal life after the war. M.I.T. needed a large number of men for routine teaching. As far as the higher administration of M.I.T. was concerned at the time, routine teaching and nothing but routine teaching was the function of the mathematics department.

There were, it is true, a few devoted spirits in the M.I.T. mathematics department who hoped against hope that the day was to come when M.I.T. would take its place beside Harvard and Princeton as a great center of creative American mathematics. These mathematicians kept bravely defying an unfavorable environment, for M.I.T. was then simply an engineering school, and mathematics was regarded as only a tool to implement engineering training. Nonetheless, they found a certain tolerant sympathy even if not much active co-operation from Professor H. W. Tyler, the head of the department. Professor Tyler, a small, active, bearded man, was not himself a research scholar and was originally quite content with his department as a service department: that is, as one contributing to the education of people whose real interests were in engineering fields. But, like every good administrator, he was ready to seize any opportunity for the advancement of his department, and later, when the chance for a certain research prestige came, he stood behind us.

Many of my colleagues were friendly, but the one who did the most to encourage me was C. L. E. Moore. He had the human gift of affection and a love for mathematics that stimulated others to reach a level to which he could never aspire himself. I wish to pay tribute here to the selflessness and the integrity of this tall, slightly awkward, humorous, and kindly man.

During my first years at M.I.T. I lived at home. My older sister, Constance, had been through Radcliffe and was engaged in graduate study in mathematics at Chicago. The accounts she gave of her rather orthodox mathematical training excited my ambition but left me very much in doubt as to whether I was really aimed toward any large success. My younger sister, Bertha, was studying chemistry, first at Radcliffe and then later at M.I.T.

I tried to develop a certain social life at the time through the Sunday teas which my parents held at their house and among my sisters' friends. In this life I remained clumsy and was still subject to the peremptory demands of my parents. They tried hard to pick my friends for me and to reject those whom they did not find suitable. In fact, they exercised a complete right of veto over the young women to whom I was paying attention. This veto was governed more by what my parents conceived to be the girls' reaction to the rest of the family than by any factor directly concerning me. It left me frustrated and confused and always more determined to make use of my summer vacations to get out from under the burden of family dominance.

The Institute of Technology was at least one place where family pressure could not easily follow me. I taught a heavy schedule, more than twenty hours a week, but I still found time not only to study mathematics but to create it. In the strength of my youth I spent the whole day at M.I.T., from nine in the morning to five in the evening, and even then I could find no greater delight than to pass my Sundays (Saturdays were working days) in my deserted office, my thoughts undisturbed by anyone. One fifth of what I did then would be too much for me now.

As to my amusements, besides the movies and the repertory company at the old Copley Theater, I went for walks in the Blue Hills or the Middlesex fells and undertook the construction of a crude toboggan with which to glide down the slopes behind the Mount Auburn cemetery. I had a few friends among my younger colleagues and the graduate students at Harvard. In the winter I used to walk on the ice to M.I.T. and to Boston from our home on Sparks Street, and in the spring and the autumn I engaged in a little mild and very incompetent tennis.

It was at M.I.T. too that my ever-growing interest in the physical aspects of mathematics began to take definite shape. The school buildings overlook the River Charles and command a never changing skyline of much beauty. The moods of the waters of the river were always delightful to watch. To me, as a mathematician and a physicist, they had another meaning as well. How could one bring to a mathematical regularity the study of the mass of ever shifting ripples and waves, for was not the highest destiny of mathematics the discovery of order among disorder? At one time the waves ran high, flecked with patches of foam, while at another they were barely noticeable ripples. Sometimes the lengths of the waves were to be measured in inches, and again they might be many yards long. What descriptive language could I use that would portray these clearly visible facts without involving me in the inextricable complexity of a complete description of the water surface? This problem of the waves was clearly one for averaging and statistics, and in this way was closely related to the Lebesgue integral, which I was studying at the time. Thus, I came to see that the mathematical tool for which I was seeking was one suitable to the description of nature, and I grew ever more aware that it was within nature itself that I must seek the language and the problems of my mathematical investigations.

Of my many older colleagues at M.I.T., one in particular helped me to become aware of the physical side of mathematics. Henry Bayard Phillips, whose active days are not over yet, is a long, lean, ageless Carolinian who grew up during a period when the bad days after the Civil War were no remote memory. He was and is the complete individualist, with more interest in doing new things than in publishing them. It was from Phillips, more than from anyone else, that I learned the importance to the pure mathematician, of a physical attitude and that I became aware of the great work of Willard Gibbs on statistical mechanics. This was an intellectual landmark in my life.

Willard Gibbs, America's greatest scientist, had himself worked in this middle ground and indeed had pioneered in it. He had lived a quiet, retired life at Yale, where he died in 1903, almost unknown to his students and colleagues. He made many contributions, both to physics and to mathematics, but the part of his work which has always been of the most interest to me was his statistical mechanics. And it is with respect to his work in this area that he exercised such a profound influence on my own career.

The great physical tradition of Newton had necessarily been one of determinism, where a perfect knowledge of the universe at one instant is understood to involve a perfect knowledge of its history throughout all time. It would have been Newton's assumption that to give the present positions and speeds of the particles in a wave moving across the surface of the Charles would allow us to plot the movement of all this wave forever. Unfortunately, no perfect knowledge of the present is available to us with our limited measuring instruments, and the problem that faces the working physicist is to find out how far he can go with the imperfect knowledge available to him.

For this he must work not with a single, fixed universe but with many different universes simultaneously, each having some preassigned probability. He cannot tell you what will always happen but what may happen at certain times, given certain conditions. This new physics of probability had been in the course of development for a considerable period, but it was toward its final clear formulation that the work of Gibbs was unquestionably tending.

When I came to M.I.T., I was intellectually prepared to be influenced by the work of Gibbs. Just before my term work had started, Dr. I. Barnett, of Cincinnati, had turned up in Cambridge and had talked various mathematical and personal matters over with me. As this was the first time that I had taken on a really mature job in mathematics, I was curious as to what problem should form the center of my new work. I asked Barnett to suggest a new and lively problem, and he mentioned to me that a lot was being done on the generalization of the concept of probability to cover probabilities where the various occurrences being studied were not to be represented by points or dots in a plane or in a space but by something of the nature of path curves in space.

For example, a one-point probability problem might concern the distributions of bullet holes in a target and ask what we can say in advance about the way they will bunch in the bull's eye. On the other hand, a curve probability problem will come up if we are asked to characterize the flight of a bee or, even better, the walk of a man who is so drunk that there is no relation at all between the direction of his previous step and his present step. If we put such a man in the center of a square field of given dimensions, how long will it take on the average for him to get out of that field?

This new concern with the probable outcome of irregular behavior has a certain historical significance. The early part of the twentieth century had seen a change in mathematics toward a more complicated world view. The great interest of the nineteenth century had been the study of points and of quantities depending on points. The new concern tried to do for curves what the older analysis had done for points.

The roots of this new emphasis were firm in the work of the nineteenth or even of the eighteenth century, which were concerned with the calculus of variations. The straightforward differential calculus of Newton and Leibnitz had discussed such maximum and minimum problems as those of the summit of a hill or the bottom of a bowl, or the related problem of the shape of a mountain pass. The calculus of variations discussed such problems as that of the quickest way of going from one point to another by a curve through a region in which the possible velocity of travel varies from point to point.

Yet, although the origins of a mathematics of maxima and minima for curves were very old, the full development of the subject was not so old. The world of curves has a richer texture than the world of points. It has been left for the twentieth century to penetrate into this full richness.

As a result of Barnett's suggestions, I spent my first year at the Massachusetts Institute of Technology investigating various extensions of Lebesgue integration to systems more complicated than those discussed by Lebesgue himself. There was a paper already written in this field by a young Frenchman named Gâteaux, who had been killed in the war. But this was only a fragment; and when I investigated it further it seemed to me to lead in the wrong direction.

There were also a number of papers that hinted at the subject by the English scholar P. J. Daniell, who was at that time teaching at the Rice Institute in Houston, Texas. Daniell's work was to me far more satisfying than that of Gâteaux, and I used it as my own model. It did not, however, specifically concern families of curves, and the work in which I followed it up to cover this new field seemed to me at the time artificial and unsatisfying.

I was an avid reader of the journals, and in particular of the *Proceedings of the London Mathematical Society*. There I saw a paper by G. I. Taylor, later to become Sir Geoffrey Taylor, concerning the theory of turbulence. This is a field of essential importance for aerodynamics and aviation, and Sir Geoffrey has for many years been a mainstay of British work in these subjects. The paper was allied to my own interests, inasmuch as the paths of air particles in turbulence are curves and the physical results of Taylor's papers involve averaging or integration over families of such curves.

I may remark that in my later visits to England I got to know Taylor very well. He represents a peculiarly English type in science: the amateur with a professional competence. He is a famous yachtsman and has the open-air appearance which belongs to a yachtsman. Indeed, one of the accomplishments of which he is most proud is that of the invention of a new type of anchor for yachts.

With Taylor's paper behind me, I came to think more and more of the physical possibilities of a theory for averages over curves. The problem of turbulence was too complicated for immediate attack, but there was a related problem which I found to be just right for the theoretical considerations of the field which I had chosen for myself. This was the problem of the Brownian motion, and it was to provide the subject of my first major mathematical work.

To understand the Brownian motion, let us imagine a pushball in a field in which a crowd is milling around. Various people in the crowd will run into the pushball and will move it about. Some will push in one direction and some in another, and the balance of pushes is likely to be tolerably even. Nevertheless, notwithstanding these balanced pushes, the fact remains that they are pushes by individual people and that their balance will be only approximate. Thus, in the course of time, the ball will wander about the field like the drunken man whom we have already mentioned and we shall have a certain irregular motion in which what happens in the future will have very little to do with what has happened in the past.

Now consider the molecules of a fluid, whether gas or liquid. These molecules will not be at rest but will have a random irregular motion like that of the people in the crowd. This motion will become more active as the temperature goes up. Let us suppose that we have introduced into this fluid a small sphere which can be pushed about by the molecules in much the way that the pushball is agitated by the crowd. If this sphere is extremely small we cannot see it, and if it is extremely large and suspended in a fluid, the collisions of the particles of the fluid with the sphere will average out sufficiently well so that no motion is observable. There is an intermediate range in which the sphere is large enough to be visible and small enough to appear under the microscope in a constant irregular motion. This agitation, which indicates the irregular movement of the molecules is known as the Brownian motion. It had first been observed by the microscopists of the eighteenth century as a universal agitation of all sufficiently small particles in the microscopic field.

Here I had a situation in which particles describe not only curves but statistical assemblages of curves. It was an ideal proving ground for my ideas concerning the Lebesgue integral in a space of curves, and it had the abundantly physical texture of the work of Gibbs. It was to this field that I had decided to apply the work that I had already done along the lines of integration theory. I met with a considerable degree of success.

The Brownian motion was nothing new as an object of study by physicists. There were fundamental papers by Einstein and Smoluchowski that covered it, but whereas these papers concerned what was happening to any given particle at a specific time, or the long-time statistics of many particles, they did not concern themselves with the mathematical properties of the curve followed by a single particle.

Here the literature was very scant, but it did include a telling comment by the French physicist Perrin in his book *Les*

Atomes, where he said in effect that the very irregular curves followed by particles in the Brownian motion led one to think of the supposed continuous non-differentiable curves of the mathematicians. He called the motion continuous because the particles never jump over a gap and non-differentiable because at no time do they seem to have a well-defined direction of movement.

In the physical Brownian motion, it is of course true that the particle is not subject to an absolutely perpetual influence resulting from the collision of the molecules but that there are short intervals of time between one collision and the next. These, however, are far too short to be observed by any ordinary methods. It therefore becomes natural to idealize the Brownian motion as if the molecules were infinitesimal in size and the collisions continuously described. It was this idealized Brownian motion which I studied, and which I found to be an excellent surrogate for the cruder properties of the true Brownian motion.

To my surprise and delight I found that the Brownian motion as thus conceived had a formal theory of a high degree of perfection and elegance. Under this theory I was able to confirm the conjecture of Perrin and to show that, except for a set of cases of probability 0, all the Brownian motions were continuous non-differentiable curves.

The papers which I wrote on this subject were, I believe, the first to disclose anything very new—combining the Lebesgue technique of integration with the physical ideas of Gibbs. They did not solve certain of the problems that were implicit in the technical justification of Gibbs's work, which were later to receive a solution in the sense of Lebesgue at the hands of Bernard Koopman, J. von Neumann, and G. D. Birkhoff. This, however, took place in 1930, when the idea that Gibbs and Lebesgue had something in common was no longer a complete novelty.

While I actually wrote my first papers on the Brownian motion, there was another phenomenon which was then coming into view, and which my work could equally be considered to describe. This was the so-called shot effect, which concerns the conduction of electric currents along a wire or through a vacuum tube in the form of a stream of discrete electrons. There is no way to create a discrete stream which will not at times lump the electrons and at other times leave the stream a little sparse. These irregularities, the shot effect, are very small, but they can be amplified to audible proportions by the use of vacuum-tube amplifiers. This tube noise or conductor noise is an important limiting effect in the use of electrical apparatus when heavily loaded.

In 1920 very little electrical apparatus was loaded to the point at which the shot effect becomes critical. However, the later development—first of broadcasting and then of radar and television—brought the shot effect to the point where it became the immediate concern of every communications engineer. This shot effect not only was similar in its origin to the Brownian motion, for it was a result of the discreteness of the universe, but had essentially the same mathematical theory.

Thus, my work on the Brownian motion became some twenty years later a vital tool for the electrical engineer. However, for a considerable period my work was as if stillborn. When at last I came to write my earliest papers on the Brownian motion— during my summer in Strasbourg, of which I shall speak in the next chapter—I did not make any great stir in the world of mathematics.

The stir that a paper makes depends not only on its inner merits but on the interests of the other workers in the field. As far as American mathematics went, Veblen and Birkhoff were the great names in the period immediately following World War I. Veblen was interested in topology, of which I have already spoken, and believed that it was his destiny to introduce this abstract field as a new American mathematics, in contrast to what he considered the effete and dying European mathematics of analysis, of the differential and integral calculus. He contributed to the birth of a valuable mathematical subject, but his concern with the health of analysis has proved to be at least premature. At any rate, I was too committed to this field to accept the mandates of the new fashion.

Birkhoff was an analyst. As I have said, he had become the unquestioned leader of American analysis and was determined to keep this position. He had persuaded himself that true analysis meant primarily those fields of dynamics in which Poincaré had worked and to which Birkhoff himself had made major contributions. For him, everything else was to be relegated to the limbo of "special problems."

Thus there was no scope for my work in the eyes of the leaders of American mathematics until many years later, when a new generation had arisen, and when the pressing needs of industry and of the war had shown that the problems which I had solved or on which I had thrown light really merited attention.

My reception in Europe was much better than it was here. Maurice Fréchet, with whom I was to spend the summer of 1920, took a mild interest in my work, which in many ways was in the spirit of his own. His younger colleague Paul Lévy had already begun to pursue related directions of thought. Taylor in England was receptive to my ideas.

My old teacher Hardy was kind to one who had been his student and offered me great encouragement at a time when it was not easy to come by that commodity. Even so, I was regarded in Europe as well as in the United States as a young man of a certain peripheral ability rather than as one of the mainstays of the coming generation.

Be that as it may, I was fully convinced of the importance of my new ideas, all the more as they organized themselves quite rapidly into a subject with a neat little formalism of its own. The stuff felt right, and even at that time it would not have surprised me to learn that it would have a considerable future. In order to understand the ramifications of my work, I had to study much more than I had already known concerning waves and vibrations or, in mathematical terms, Fourier series, the Fourier integral, and the like. I began to make myself thoroughly familiar with those branches of mathematics which had already proved to be of physical significance. All this dovetailed into the desire of my colleagues at M.I.T. that I interest myself in applied mathematics. From this time on, my work was never random and desultory but had a definite direction in which I could naturally proceed.

Mathematics is very largely a young man's game. It is the athleticism of the intellect, making demands which can be satisfied to the full only when there is youth and strength. After one or two promising papers, many young mathematicians who have shown signs of ability sink into that very same limbo which surrounds yesterday's sports heroes.

Yet it is not bearable to contemplate a brief distinction and burgeoning of activity which is to be followed by a lifetime of boredom. If the career of a mathematician is to be anything but an anti-climactic one, he must devote this brief springtime of top creative ability to the discovery of new fields and new problems, of such richness and compelling character that he can scarcely exhaust them in his lifetime. It has been my good fortune that the problems which excited me as a youth, and which I did a considerable amount to initiate, still do not seem to have lost their power to make maximum demands on me in my sixtieth year.

Do not think for a moment that my new success made a hero

of me in my own home. My father was gratified by my industry and my apparent ability to produce work which at least pleased myself; but at that time my own self-laudatory claims did not receive much of an echo from his colleagues. My father was no longer sufficiently active and interested in mathematics to judge my work on its own merits.

2

The
International
Mathematical Congress
of 1920
at Strasbourg

In the present chapter as well as in those to come I shall have occasion to write about my various visits abroad. These visits represent an essential part of both my personal and my scientific life. They have not by any means been mere junkets, or interludes which have concerned me solely for my amusement, although they have in fact been very gratifying to me. Let me explain something of what they have meant to me, and in particular what was the significance of my 1920 trip.

In this, as well as in all other matters concerning my history, it is necessary to go back to my father. My father's formal education had been entirely European, and although he had gone to the Gymnasium and for a brief period to a medical school in Russian Poland, he was quite aware that his chief educational affiliations had been German. The high level of the

German Gymnasium, or classical secondary school, had led to its dominance throughout central and eastern Europe. All educated eastern Europeans had something of a German education.

But in addition, Father had a particular connection with German education. His own father had been a journalist with the Yiddish press but was a great enthusiast for the pure German language, which he vastly preferred to Yiddish. One day he decided to change the language of his journal, which was published in Byelostok, from Yiddish to High German. Naturally this cost him almost all his subscribers, and from that time on my grandfather's career had been uniformly unsuccessful.

Owing to these circumstances, my father had been brought up with literary German as his home language. Later on, he went to Berlin for a few months of engineering training at the old Berlin Institute of Technology, at a time at which this school had not yet moved from the center of Berlin to its famous later site in Charlottenburg. Father's academic training there was brief and stormy. He ultimately left Germany (of which country he was a subject) to engage in a nebulous scheme for founding a humanitarian vegetarian colony somewhere in Central America.

The scheme was never well conceived. The other young man who was to come over with Father soon found that he had no stomach for the adventure. Father was stranded, penniless, in New Orleans, and never got to Central America at all. For some years he led a Huckleberry Finn existence in the West and in the South.

Finally he found it possible to return by a roundabout path to the career for which he was most fitted: that of linguistics. He taught for a period at the Kansas City Central High School and later at the University of Missouri, from which he came on a venture to Harvard and attracted the attention of Professor Francis Child, the editor of *Scottish Ballads*. This led to an instructorship at Harvard and finally, after many years, to the chair of Slavic Languages and Literatures.

Father was an enthusiastic scholar, with interests spreading well beyond the field of Slavic languages. He had fancied himself as a great scholar after the German pattern. This caused a considerable ambivalence in him.

On the one hand, he was essentially a German liberal of a type well known in the middle and later nineteenth century, fully in sympathy with the German intellectual tradition as it had come down from the time when Goethe had been a truer symbol of German aspirations than the Emperor Wilhelm II ever was to become. Separated as my father was from Germany, largely self-educated, and outside the orthodox German academic tradition, he still hoped for many years that by sheer intellectual strength and integrity he could win from Germany the recognition accorded to a great German scholar.

In this expectation Father had never been completely realistic. He was too innocently honest a man to be worldly-wise. It took him many years to learn how the great German intellectual tradition had come to be subservient to a group of vested interests in learning. The really great advances of the German Empire after the Franco-Prussian War had brought with them a not inconsiderable degree of the spirit of the climber and the worshiper of material success. I must admit there were many men in the German universities who were willing to accept new ideas from any source whatever. For all this, as the years went on, the entry to the inner circles of German intellectual life became more and more difficult for the outsider to achieve. This was particularly true in fields like philology and linguistics, in which a definite decision as to the merit of new work is difficult or impossible. The doors of complete recognition were closed to my father.

When Father visited Germany about the beginning of the century, and again later in 1914, he found himself far more of an outsider than he had expected or even feared. This hurt him. He came to resent and indeed to hate Germany, with the sort of hatred which one reserves for those kinsmen who, one feels, have let one down unjustly.

This hatred was reinforced by the political and social changes he had observed in the new Germany.

In addition to that, he did not like what he had seen of German militarism just before the First World War. He came to be one of the chief American supporters of the point of view of the Allies—of France and England. Together with Professor Bierwirth of the German department of Harvard, another ardent anti-militarist, he would walk down Brattle Street every morning berating Germany in the German language. The very intensity of his emotion marked a personal involvement in European matters which was entirely different from the potential isolationism of the average American and even of the average American scholar.

My father was a great enthusiast for America and for most of what was American, but at the same time an enthusiast from outside. He was deeply critical of much in America, and in particular of the shallowness of a great part of American education. The explicit intensity of his love for America and the specific character of that love were perhaps the least American things about him. He loved America, that is, as if it had been his personal discovery, rather than a background so close to him that he could take it for granted.

We had been in the habit of entertaining at our house a number of European scholars, who were largely liberal and disaffected toward the actual state of affairs in Europe. Some of them were among the great reformers of the beginning of the twentieth century. We had as house guests, for instance, Thomas Masaryk, later to become president of Czechoslovakia and the greatest elder statesman of Europe; Paul Milyukov, historian, economist, member of the Russian Duma and finally associate of Kerensky; Father Palmieri of the Propaganda, the chief Catholic authority on the Eastern Church and its various Uniat churches which had drifted into the orbit of Catholicism; and, during the World War itself, Michael Yatsevich, the Siberian engineer to whose lot it had fallen to draw up many contracts for Imperial Russia and to secure the moneys arising from them against the claims of the Communists, for some future non-Communistic and possibly democratic Russia.

Thus it was a familiar part of our life to hear foreign languages spoken in the household. My father, indeed, could speak some forty of them. He was so proficient in linguistic matters that his insistence as a teacher on accuracy and fluency had the somewhat surprising effect of almost completely inhibiting the efforts of my mother and of us children to speak more than one language.

With this background it was inevitable that I should develop a great curiosity about Europe and a profound thirst for the springs of European scholarship. These factors were supplemented by others pertaining more directly to myself and to my own interests. My years in England and, to a slighter extent, my term at Göttingen had constituted almost the first true relief I had found from the intensity of our home life and my parental pressure. My earlier research training had been primarily English and, to a lesser extent, German. The friendly recognition which I had already begun to find in Europe contrasted sharply with the sense of rejection which I had experienced around Harvard.

At the English universities, it is true, there is a certain gentleman's pretense that one is an amateur and not too deeply concerned in the hard and grinding work of scholarship. This was generally understood to be a pose. It did not require any great insight to see that those very men who conformed to the convention of the pococurante were tremendously excited about ideas and more than eager to talk about them. At Harvard, on the other hand, the pose of a lack of interest in creative scholarly work was more than a convention. The typical Harvard man considered it bad taste to talk too much and to think too much about science. The effort of trying to be a gentleman was enough of a strain on his resources.

It is thus easy to understand how the end of the war found me starved for European contacts and only too eager to return to the relative freedom of a European trip. Here the strong hand of the family could scarcely reach me. But there was still another reason why the trip tempted me: the coming International Mathematical Congress at Strasbourg.

Normally, in all fields of science, it has been the custom, say every four years, for those working together in a major subject like mathematics or physics or chemistry to foregather in some central place, read papers, and talk over the problems of their work. The first great war (alas!) interrupted these indications of the universal humanity of science, and the present alignment of the world into two hostile camps has tended to frustrate these meetings still further.

The last International Mathematical Congress before the war had taken place in England in 1912, at Cambridge. The congress which was to have taken place in 1916 was clearly impossible and was allowed to go by the board. The next one, in 1920, did not find any adequate machinery established for its organization. France decided to step into the gap and celebrate an international congress in the newly re-Gallicized city of Strasbourg and at its university, now French. This had become the second university of France and the only provincial university with a great tradition of its own.

In many ways this was an unfortunate decision. It was one which later led me to regret my little share in sanctioning the meeting by my presence. The Germans were excluded as a sort of punitive measure. In my mature, considered opinion, punitive measures are out of place in international scientific relations. Perhaps it would have been impossible to hold a truly international meeting for another couple of years, but this delay would have been preferable to what actually did take place, the nationalization of a truly international institution. All that I can say for myself is that I was young and that I did not feel myself in a position of direct personal responsibility for the course taken by international science. I was avid to seize the opportunity to revisit Europe with a certain small scientific status.

I hoped that I would be able to use the period before the opening of the Congress in September in working with some European scholar in whose field I was interested. The scholar I chose was Maurice Fréchet. It was Fréchet more than anyone else who had seen what was implied in the new mathematics of curves rather than of points, the field of which I have spoken in the last chapter. At that time we all had great expectations that his work was to mark the next great step towards the mathematics of the future.

Let me say that at the present time Fréchet's work has proved to be very important but less central than we had expected. It is written in a spirit of abstract formalism which is fundamentally hostile to any deep physical application. However, hindsight is easier than foresight, and it would have been somewhat difficult when we were at Strasbourg to predict that Fréchet was not to be the absolute leader of the mathematicians of his generation.

One of the specific things which attracted me to Fréchet was that the spirit of his work was closely akin to the work I had tried to do at Columbia on topology. My training with Russell and my later contact with the work of Whitehead had sensitized me to the use of formal logical tools in mathematics, and there was much in Fréchet's work which was suited from the very beginning to be embodied in the peculiar and highly original mathematico-logical language which Whitehead and Russell had devised for the *Principia Mathematica.*

But in order for me to describe the main events at Strasbourg in 1920 I must first write in some detail about the terms "postulationalism" and "constructionalism." This dichotomy of method is one of the central issues of modern mathematics and was of great concern to me at the Strasbourg conference.

The geometry of the Greeks went back to certain initial assumptions, known variously as axioms or postulates, which were conceived to be unbreakable rules of logical and geometrical thought. Some of these were of a predominantly formal and logical character, such as the axiom that quantities equal to the same quantity must be equal to each other. Another, with a more purely spatial content, was that known as the parallel axiom, which asserts that if we have a plane containing a line l and a point P not on that line, then through P and in that plane one and only one line can be drawn that will not intersect l. This will, of course, be the line parallel to l.

This postulate does not have the simple obviousness of the purely logical postulates of mathematics. And generations of mathematicians have sought exceptions to it. In the eighteenth century the Italian mathematician Saccheri spent a considerable amount of effort to ring the changes on the parallel postulate with the hope that any denial of this assumption would have to lead sooner or later to a logical contradiction. He did a brilliant job with the various modifications of the axiom, but his effort was unsuccessful. In fact, the more he tried to draw a contradiction out of the denial of the postulate, the larger was the body of doctrine which he was able to derive from this denial. In fact, this body of doctrine amounted to a geometry which was essentially different from the usual geometry of Euclid, but which was rather grotesque than self-contradictory.

Finally, in the beginning of the nineteenth century, a group of mathematicians including John Bolyai, in Hungary; Lobachevski, in Russia; and ultimately the great Gauss, in Germany, came to the bold conclusion that there was no contradiction in denying the parallel postulate, but merely a new and different, non-Euclidean geometry. From that time on it came to be more and more clearly understood that the so-called postulates of geometry, and indeed of other mathematical sciences, were not undeniable truths. They came to be regarded as assumptions which we could make or refuse to make concerning the particular mathematical systems which we wished to study further.

This tentative attitude in mathematics, in which the postulates became suppositions made for the sake of further work rather than fundamental principles of thought, gradually began to be the standard point of view of mathematicians in all countries. In America one of its earlier exponents, and perhaps its chief early exponent, was Edward Vermilye Huntington, of Harvard, with whom I had studied in 1912, and who had exerted a great influence on my ways of thought.

Whitehead had been perhaps the chief English postulationalist, but he supplemented a pure postulationalism with the view that the objects of mathematics were logical constructions rather than simply the original concepts described in the postulates. For example, at times he regarded a point as the set of all convex regions which in our ordinary language might be considered to contain this point. As a matter of fact, Huntington has formulated very similar ideas quite independently, and an important essay in this direction had been made by the philosopher Josiah Royce several years earlier. But the classical example of constructionalism in mathematics is the definition of the whole numbers which occurs in the *Principia Mathematica* of Whitehead and Russell.

The distinction between the postulationalist treatment of numbers and the constructionalist treatment described by

Whitehead and Russell is that in postulationalism the numbers are undefined objects which are connected by a set of assumed formal relations and specific things which can be built up out of our experience by definite modes of combination of even more elementary experience. A postulationalist treatment of numbers makes them simply objects to be arranged in a before-and-after relation so that if *a* is before *b* and *b* is before *c*, *a* is before *c*, and every number other than o has another number immediately before it, and so on. These are some of the postulates in such a treatment of the number system.

In the constructionalist treatment of numbers, a unit set is taken as a set of entities all of which are the same. The number 1 then becomes the set of all unit sets. A diad is a set of entities which is not a unit set but which becomes a unit set on the removal of any one of its component entities. Two is now the set of all diads. A triad is a set of entities which is neither a unit set nor a diad but which becomes a diad when any one entity is removed. Three is the set of all triads. In this way, and by the process known as mathematical induction, the complete set of all the natural numbers can be built up.

To the layman all of this will perhaps sound rather like empty logic chopping. Have we not in fact used the numbers 1, 2, and 3 in only a slightly obscured form when we built up these definitions of the earlier whole numbers? But to the logician this objection does not seem cogent for the greater precision of thought given to these definitions furnishes him with a firm ground on which to stand and from which to advance to further mathematical ideas.

The trick of building up more and more complicated mathematical objects as sets of sets and relations between relations had become familiar to me both from the work of Huntington and from that of Russell. I had indeed already written two or three essays in the application of this technique to the construction of certain elementary mathematical situations.

Postulationalism and constructionalism, as I have so far described them, do not represent movements only in mathematics. Postulationalism in particular is shared by physics. The relativity of Einstein and the new quantum mechanics both constitute regions in which physics has burst the frame of classical Euclidean geometry and assumed new definitions which are given rather as sets of explicit axioms than as a rigid and irreplaceable spatial intuition, as the older Kantian theory of space demanded.

It is true of course that the tendency to postulate for the sake of postulating and to write papers for the sake of writing papers characterizes a considerable amount of the newer mathematics. Nevertheless, the cold, hard medium of logic, like the cold hard medium of marble, imposes a certain internal discipline on all but the most vacuous and trivial mathematicians, even when they favor the newer mode of freedom.

As I have said, I was brought up in the postulationalist tradition and shared in the early years of the constructionalist tradition which arose therefrom. When I looked for a French scholar with whom to study, I looked for one whose work might embody one or both of these two directions of thought. As far as these demands were concerned, Fréchet had no rival among French mathematicians.

So far I have spoken of the new direction of thought largely from the Anglo-American point of view. In Germany, too, there had been some early exponents both of postulationalism and of constructionalism, and here the leading and most original names were those of G. Frege and Schröder. France, on the other hand, was rather late to adopt these newer habits of thought, but, as far as France had gone in postulationalism, Fréchet was the unquestioned leader. I myself had made one or two not completely unsuccessful attempts to supplement Fréchet's postulationalism as a tool for the study of new and more complicated spaces of curves by means of a constructionalist attitude. This new effort was, however, outside the inner frame of Fréchet's own work.

I wrote to Fréchet to find out if he was willing to take me on as a disciple during the summer vacation preceding the Strasbourg congress, and I received in return a cordial letter of invitation from him. His first plans had been to spend his holidays in Béarn, on the Spanish border. Later, however, he changed his mind and invited me to come and work with him, first in Strasbourg, and then in a little country village in Lorraine which has the German name of Dagsburg and the French name of Dabo.

Early in July, I embarked for France in the French Line steamer *La Touraine* with some family acquaintances who had promised my parents to keep an eye on me during the voyage. However, the gin-soaked twenties were already upon us, and I found that my friends' ideas of what was becoming on an ocean voyage were incompatible with my rather puritanical personal habits.

I had never been a teetotaler. I enjoyed the wine we had with our meals on the boat, which I diluted plentifully with water. On the other hand, I did not like distilled liquors, and I protested vigorously against the convention under which my acquaintances pressed me to drink. I find this habit of urging drink upon a person who manifestly does not want it, quite as much an infringement of personal liberty as any blue-nosed prohibitionism can be. Thus, I was not happy on the boat and I made no friends. I was eager to land and to get away from my companions of the voyage.

But before we landed we had an interesting and not altogether pleasant experience of another sort. It had been an overcast voyage, and we were unable to shoot the sun for several days. We were going full speed ahead by dead reckoning, for these were the days in which wireless was simply a means of communication rather than an aid to navigation through an accurate system of cross bearings. We were due to make our landfall at Bishop Rock when suddenly we saw rocks looming from the fog all about us. The engines were set full speed astern, but not before we got into difficult waters.

While backing out we ripped a considerable hole in our stern. Water came in to the steerage compartment. I am told that there were the beginnings of a panic. This was controlled by the ship's officers together with a French prize-fighter passenger whose prestige was enough to keep order.

We were all sent down to our quarters to put on our life belts. It was not pleasant coming up on deck again in the midst of the milling crowd. I felt that I would like to be on deck as quickly as possible, but I also felt that any attempt to rush the matter or to get ahead of other people would not only be an act of cowardice but an act of treason to our common welfare. I forced myself to make way for others and to go up at a measured pace.

When we came on deck we still did not know what awaited us. The ship was taking in water, and the bulkheads threatened to give way. The ship's carpenter did yeoman work reinforcing them so that they did in fact hold throughout our trip across the channel to Le Havre. But still we were ordered to spend the night on deck, sleeping in our life belts by our boat stations. I remember that somebody dropped a bottle by my head as I slept.

We disembarked ... Le Havre on the next morning without further incident, b< . the boat had been damaged far beyond our knowledge and was taken out of service for many months. There was mail awaiting me on shore. I learned that Fréchet was not quite ready to receive me yet, so after a few hours I recrossed the Channel to Southampton and went up to Cambridge.

At Cambridge I found some old friends waiting for me. I stayed with Dr. and Mrs. Bernard Muscio, a couple of Australian psychologists. I had known them in my student days at Cambridge, and they had visited Boston a few years before on an English war mission. I looked up a number of other acquaintances, including Hardy, who was about to leave Cambridge for a chair at Oxford.

In general I found that I had not been forgotten at Cambridge and that my colleagues were glad to give me that cordial reception which I had never received at Harvard. I had not been matriculated at Cambridge, as I had gone there on the basis of a special arrangement with Harvard University, permitting me to take courses without that formality. Years afterward I asked Jessie Whitehead, the daughter of Alfred North Whitehead, whether I could consider myself a Cambridge man. Under the circumstances, she said, the best that I could call myself was an illegitimate son of Alma Mater. At any rate, I have found Alma Mater more than ready to accept her by-blow into the family.

After a few days I left for Paris, where I stayed in a cheap hotel with unbelievably bad sanitary accommodations near the Louvre. I did not find my vegetarianism much in my way in France, for there was a wealth of cheap restaurants with good and appetizing vegetable dishes. I had no friends in Paris, nor was my French at that time more than barely usable. Moreover, the Paris café and street life shocked my youthful puritanism, and I was profoundly homesick and unhappy.

It seemed to me as if the house doors of a great city were a serried wall of fortifications, impregnable to the foreigner. I spent my abundant leisure walking the streets of the city and visiting the museums, particularly one tidbit to which a French-trained friend in America had alerted me: the Museum of the Ecole Centrale des Arts-et-Métiers, in which the remains of the inventions of the nineteenth century and the apparatus used for great scientific experiments were conserved in a peculiarly French sort of dusty disorder.

Fréchet had made an appointment with me, first at a *lycée* on the Boulevard Saint-Michel, where he was grading examinations, and later for lunch at an Alsatian *brasserie* on the same boulevard. He was a mustached, sinewy, athletic man of medium height. He had served in the French army as an interpreter for the British, and he was as enthusiastic for walks and tours as I was. We hit it off well together. He was, however, still not ready to receive me at once in Strasbourg, so I made a little detour into Belgium to visit some friends before I should settle down for work with Fréchet. I found my friends in their fine old house in Louvain, which had just been restored from the disorder and filth in which it had been left by the German officers who occupied it during the war, but my trip was badly timed, as it coincided with a visit of President A. L. Lowell of Harvard and his wife. Thus my entertainment was largely a matter for the children, and in particular for a young son of the family who had just completed a year at the Harvard Law School. He took me about the burned and ruined town and showed me the remains of the library, the town hall, and the nave of the church, half blocked with scaffolding. He also took me on various walks about the countryside and gave me his confidences.

Now that he was free of the Harvard atmosphere, he felt at liberty to protest against certain matters of English and American legal education. He did not like the case system and greatly preferred the a priori way in which those countries deriving their jurisprudence from Roman law search for legal principles rather than for legal precedents.

A few days later I traveled to Strasbourg by way of Luxembourg and the iron country. It was a relief for me to find myself in a region where German was spoken more freely than French, for German was a language at which I was decidedly more proficient. I settled in a boarding house in a new part of the city. Every day or every other day at the least I had a few hours' consultation with Fréchet in the garden of his little house beside the Ill–Rhine Canal.

There were two or three points in Fréchet's work which I tried to extend. Fréchet's treatment of more generalized spaces had made no use of what we know as co-ordinates: that is, he had made no attempt to represent his points by sets of numbers. In the co-ordinate representation of space, any two points at the ends of a line interval are measured by the difference between corresponding numbers at one terminus and at the other. In ordinary geometry of two or of three dimensions, this method of representing a line segment is known as its vector representation. For example, given one point in three-dimensional space, I can locate another with respect to it by saying how far I have to go north from the first point, how far east, and how far up, to get to the second point.

The theory of vectors is not new in mathematics. For well over a century and a half it has been known that an ordinary space of three dimensions contains in itself directed quantities like arrows, which can be added to one another, as for example by traversing the step indicated by one arrow and then the subsequent step indicated by another and by considering this double step as if it were a single step. It is beside the point here to go into the many operations which can be conducted with these directed steps, but it has long been known that similar geometries exist in spaces of more than three dimensions and, in fact, in spaces of an infinity of dimensions.

Fréchet's generalized theory of limits and of differentials applies to many sorts of spaces, including vector spaces, but is not necessarily confined to those spaces in which the elements may be regarded as steps. On the other hand, this geometry of steps constitutes a very important part of Fréchet's general theory and was worth solidifying with an appropriate set of postulates. Fréchet had not done this, nor did he consider these particular vector systems as peculiarly important among those

which he had considered. This was the task which I had performed. It was very closely allied to the theory of the combination of successive transformations which is known as group
theory, and in fact it constitutes a significant chapter of
that theory.

However, I gave a full set of axioms for vector spaces.
Fréchet liked it, but did not seem particularly struck with the
result. But then, a few weeks later, he became quite excited
when he saw an article published by Stefan Banach in a Polish
mathematical journal which contained results practically identical with those I had given, neither more nor less general.
Banach's conception of his ideas and his publication of them
were both a few months earlier than my own. There had, however, been no chance for communication between us, and the
degree of originality of the two papers was identical.

Thus the two pieces of work, Banach's and my own, came for
a time to be known as the theory of Banach-Wiener spaces. For
thirty-four years it has remained a popular direction of work.
Although many papers have been written on it, only now is it
beginning to develop its full effectiveness as a scientific method.

For a short while I kept publishing a paper or two on this
topic, but I gradually left the field. At present these spaces are
quite justly named after Banach alone.

There were several motives which led me to abandon this
brain child, of whom I was at least one of the parents. The first
was that I did not like to be hurried or to watch the literature
day by day in order to be sure that neither Banach nor one of
his Polish followers had published some important result before
me. All mathematical work is done under sufficient pressure,
and its increase by such a fortuitous competitive element is
intolerable to me.

But the important reasons for my accepting or rejecting any
specific piece of work have to do with the much-neglected field
of mathematical aesthetics. To ask exactly what I mean by this

poses a peculiarly difficult problem, for I must convey to the non-mathematician not only the substance and the texture of the mathematical work I have done but my emotional responses to it. I shall have to tell him my reasons for rejecting certain problems which have proved interesting to other people for a considerable number of years but do not seem to me to offer the same opportunities for my own vein of work and for the display of my own mathematical taste and for my particular individual powers.

This leads to a problem which must be faced in one form or other by any autobiographer who has done significant work in a difficult and private field like mathematics. A composer cannot avoid paying a certain amount of attention to the techniques of composition, and to aspects of harmony and counterpoint which are the very substance of his work but which can be appreciated only in a very limited degree even by the devoted concertgoer who has not himself faced the task of musical composition. The writer or the painter is no less involved in this problem if he goes in for autobiography. He may seem to address himself to the educated layman who can appreciate the results of his creative work. However, he has not completely performed the task of the autobiographer unless at the same time he has managed to express himself concerning those tasks of writing and of painting which can be fully appreciated only by the man who himself has faced them on a high professional level.

But the creative artist and even the creative musician are better able to attract the attention of the lay reader than the mathematician. The layman is easily convinced that, whether he is himself immersed in these creative tasks or not, a degree of information concerning the nature of their creativity belongs to his status as a cultured man. He is further aided by the fact that, even for the reader who makes no claim to the understanding of the full technical means by which a specific emotional effect is reached, there is a direct appreciation of the emotional effects of the arts, sufficient to lead the layman to a sincere effort to understand how the artist arrives in detail at beauties which the appreciative observer may see only as an accomplished whole.

The specific difficulty of the mathematician when he undertakes autobiography is that the layman does not conceive it to be any part of his aesthetic and cultural duty to understand the least thing about mathematics. For him, mathematics is likely to be a dull, dry, and formal subject. If the general public ever thinks of mathematics, it sees it at best as a tool for the physicist and the statistician and at worst as something closely akin to the work of the accountant. Hardly any non-mathematician will admit that mathematics has a cultural and aesthetic appeal, that it has anything to do with beauty or power or emotion.

I categorically deny this cold and rigid concept of mathematics. A piece of mathematics may have the virtues of logic and rigor yet, in the technical opinion of the trained observer, it may be insensitive and purely formal. To other mathematicians, the task of the mathematician is to use a rigid and demanding medium to express a new and significant vision of some aspect of the universe; to express *apercus* which reveal something new and something exciting. If his medium is strict and confining, so are in fact the media of all creative artists. The counterpoint of the musician does not interfere with his perceptivity, nor is a poet less free because his language has a grammar or his sonnets a form. To be free to do anything whatever is to be free to do nothing.

What differentiates the appeal of the artist-mathematician from the artist-sculptor and the artist-musician is not the unemotionality of his public but the strict discipline necessary to become a connoisseur of mathematics. It is quite possible to imagine a community of musical composers whose primary satisfaction will arise from the interchange of the musical scores they have composed. They might well be relatively indifferent to the performance of these scores at concerts attended by those capable of understanding them only through the vaguer channels of receptive emotional feeling.

That the mathematician displays this aloofness from his public is indeed due not so much to an intellectual-aesthetic snobbery as to the very high degree of training at which it is necessary for the amateur to arrive in order to acquire even an appreciative relation to the content which is presented him and to the fact that, short of this technical appreciation, there seems to be no channel through which the layman can get to feel anything at all, even passively.

This limitation is not as absolute as it may seem at first sight. There is actually a considerable and growing public of trained engineers and natural scientists who, while their first interest may have originally been in the use and application of mathematics to purely utilitarian ends, have nevertheless acquired a sufficient background to appreciate a powerful theory or a clever and elegant proof. At least part of my motive in writing this book is to call the attention of the public at large to the existence of this more limited public of amateur mathematicians. I also mean to give to the reader outside this small circle at least a hint of the thrill of mathematical creation.

It is thus in an aesthetic rather than in any strictly logical sense that, in those years after Strasbourg, Banach space did not seem to have the physical and mathematical texture I wanted for a theory on which I was to stake a large part of my future reputation. Nowadays it seems to me that some aspects of the theory of Banach space are taking on a sufficiently rich texture and have been endowed with a sufficiently unobvious body of theorems to come closer to satisfying me in these respects.

At that time, however, the theory seemed to me to contain

for the immediate future nothing but some decades of rather formal and thin work. By this I do not mean to reproach the work of Banach himself but rather that of the many inferior writers, hungry for easy doctors' theses, who were drawn to it. As I foresaw, it was this class of writers that was first attracted to the theory of Banach spaces.

The chief factor which led me to abandon the theory of Banach spaces, after a few desultory papers, was that my work on the Brownian motion was now coming to a head. Differential space, the space of the Brownian motion, is itself in fact a sort of vector space, very close to the Banach spaces, and it presented itself as a successful rival for my attentions because it had a physical character most gratifying to me. In addition, it was wholly mine in its purely mathematical aspects, whereas I was only a junior partner in the theory of Banach spaces.

I do not believe that Fréchet appreciated the importance of differential space when I first mentioned the theory to him. However, he got me in touch with Paul Lévy, of the Ecole Polytechnique, then the most promising young probability-theory man in France, whose work and mine have exerted a mutual influence on each other up to the present time. It cost me a little effort to persuade Lévy that my work was essentially different from that of Gâteaux, but he soon saw the point. He was to become one of my closest friends and supporters.

Curiously enough, another colleague whose work has always been in the closest relation with Lévy's and mine was the Swedish mathematician Cramer, whom I had met during my stay in England as another house guest at the Muscios' that summer.

For some time my mathematical interests made me oblivious to my personal comfort. When I came to sufficiently to think of my environment, I found that I was lonely at my boarding-house. There was an American there who had come over with me on the same boat and who had no high opinion of me. A young English composer at the same boardinghouse was a friend of his. I wanted this composer to like me, but I had started off on the wrong foot. My fellow passenger of *La Touraine* certainly did nothing to help my standing in our little community.

The musician regarded me as heavy-handed and Philistine. This was partly because of my actual social ineptitude and bad manners, but it was also due to the fact that he considered that mathematics by its own nature stood in direct opposition to the arts. On the other hand, I maintained the thesis of this book: that mathematics is essentially one of the arts; and I ding-donged on this theme far too much for the patience of a man initially disposed to hate mathematics for its own sake. Later on we got into an explicit quarrel, in which we really said the unpleasant things we thought of one another, and this finally cleared up into a certain degree of understanding and even of a limited friendship.

The time came when I was to accompany the Fréchet family to Dabo. They stayed at the best hotel, and I naturally went to the other inn, in order not to infringe upon their privacy. I had ample opportunity for lonely walks over the countryside, climbing up the red sandstone hills of the Vosges and dipping down into the steep valleys, their streams drained for irrigation.

The landlord and landlady of my inn were very considerate of me. I did occasional jobs of wood chopping and the like for them. I felt at home in the countryside, among the crowing of the cocks and the lowing of the cattle. I enjoyed the sound of the water trickling down the village street to the place where the women washed their clothes, and the rhythmic beat of the flails on the threshing floor.

When finally we returned to Strasbourg, the International Congress was near. I had time in plenty to waste, exploring the quaint streets about the cathedral and following the circle of inner canals that surrounds the center of the city.

Three young American friends of mine came for the meeting. We put them up in my boardinghouse, two to a room. One was Forrest Murray of Harvard, vague and amiable, with whom I had often played tennis, and who had been a friend of our family for many years. Joe Walsh accompanied him. Joe, who was about my age, is still a professor in the Harvard mathematics department. He is tall and genial, and in those days his blond hair stood almost erect on his high forehead. He seemed to enjoy his visit to France thoroughly. He intended to stay in France for a year of postdoctoral study. His deep, booming voice was agreeable to hear, and it was pleasant, too, to participate in his eagerness for new experiences.

The fourth of our group was James S. Taylor, then like myself a new postwar section man at M.I.T., and now for many years professor at the University of Pittsburgh. Taylor was a kinsman of Phineas T. Barnum and was himself an enthusiastic showman of parlor tricks. The four of us have gone very different ways since, but at that time we were all united by youth, and we were tasting this youth to the full.

The congress guests began to roll in. From America there came Eisenhart from Princeton, with his beautiful young wife; Leonard Eugene Dickson, the number theorist of Chicago, famous as an enthusiast for France and the French, and a past master at bridge; and Solomon Lefschetz of Kansas, who had conquered the effects of a terrible industrial accident which he had suffered as engineer with the Westinghouse Company at Pittsburgh and had entered upon the new career in mathematics which was to take him to the leadership of the mathematics department at Princeton University and to the presidency of the American Mathematical Society.

There were several of the older people at the meeting whom we felt as links connecting us with the great past of mathematics. Sturdy old Sir George Greenhill represented Woolwich. Camille Jordan, who for all his ninety years accompanied us on our pedestrian excursions, was like a memory from the days of Louis Philippe. His recollections dated back to the great days when Cauchy was lording it over French mathematics and forcing all the younger men to pay tribute. When Jordan died, two years later, we all felt his death as a break in the continuity of the mathematical tradition.

Professor Jacques Hadamard, of Paris, played a great part at the congress. He was then only in his middle fifties, but his reputation had been well established before the end of the nineteenth century, and to us fledglings he was a great historical landmark. Small, bearded, very Jewish-looking in the *fin* French way, he occupied a unique position in the affections of his younger colleagues.

English mathematics belongs to Oxford and Cambridge, where there are ample bonds between the undergraduate and the don; while German mathematics is characterized by the amenity of the *Nachsitzung*. The official discussion of a scientific paper is followed by a procession across the university town to a beer garden, where the great and the little alike talk over the latest results in mathematics as well as the trivial pleasures of life. French mathematics, however, has followed a largely official course, and when the professor has retired to his little office and has signed the daybook which gives a record of the lecture he has just finished, it is customary for him to vanish from the lives of his students and younger colleagues.

To this withdrawn existence, Hadamard forms an exception, for he is genuinely interested in his students and has always been accessible to them. He has considered it an important part of his duty to promote their careers. Under his personal influence, the present generation of French mathematicians, for all the tradition of a barrier between the younger and the older men, has gone far to break down this barrier.

I myself benefited from Hadamard's largeness of outlook. There was no reason why Hadamard should have paid any particular attention to a young barbarian from across the Atlantic, just at the beginning of his career. That is, there was no reason except Hadamard's good nature and his desire to uncover mathematical ability wherever he could find a hint of it.

Many years later, when I was to meet him again at various mathematical congresses and as a fellow lecturer in China, I was surprised and gratified to find that he still remembered me and had an accurate idea of the entire development of my work. Thus one very positive result of the Strasbourg meeting was to bring me together with the long succession of French mathematicians who owe their recognition and their careers to Hadamard.

We congressionists made a number of interesting excursions around the sights of Strasbourg—to Saverne and the great ruined keep of Haut-Barr, to the quaint old city of Colmar, and to a sector of a battlefield of World War I. The French soldiers took us there in army trucks, but on the way back the trucks broke down and we had a long and tedious wait. Thus, when we came to the little wine-growing village of Turckheim, where the meal that had long been waiting for us had got cold and been abandoned, the mayor presented each of us with two glasses of the wine of the country, one old and one new. We had to sip these as a matter of courtesy. I will not deny that the wine was excellent, but two glasses on an empty stomach represent a severe ordeal, and new wine has a potency of its own. As we continued to Strasbourg by truck and by train, some staid souls were a bit less inhibited than was their wont.

Soon the meeting was over and the four of us went back to Paris. Taylor was to return to America with me, but my other two American friends had decided to spend a year studying in Paris. They were eager to immerse themselves in a thoroughly French environment, and they made it quite clear that we should confer on them no favor by continuing to forgather with them.

The remaining two of us found ourselves without return accommodations. *La Touraine,* on which I had come to France and on which we had intended to return, had not yet recovered from its near shipwreck, so all through a very autumnal September and part of October we waited in vain for the resumption of service, believing to the very last that we should be back for the beginning of the M.I.T. term. This however was impossible.

We began to worry what the school might be thinking of our tardiness. Nonetheless, we enjoyed ourselves in Paris and walked over large areas of the town for lack of anything else to do. We haunted the shipping agencies and finally heard of a new American boat just about to be put into transatlantic service. The passenger list was small and interesting and consisted largely of old globe-trotters. The travelers Osa Johnson and her husband were on board, together with a tame orangutan that they had brought from Indonesia. This orangutan contrasted pleasantly in its cultured behavior with a demon child who insisted on scattering the chessmen with which we were playing in the smoking room.

I came back from Europe with renewed and enlarged inspiration. For all of the defects of my French, I had lived in France and had for the first time established a contact with my French colleagues. Both in France and in England I had found that I occupied a more important position than at home, and I had work pending which seemed (at least to myself, if to few others), the promising beginning of a career in mathematics.

I had indeed seen the physical devastation of war, both in Belgium and along the former fighting line in Alsace. I had, however, become aware that the European spirit had greater possibilities of recuperation than I had previously assumed. Notwithstanding the war in the West, the defection of Russia from the European camp, and the news of battles in Poland, it was still possible to hope that World War I was merely an interlude and would be succeeded by another period of peace as long as the great peace of the nineteenth century.

As to Russia and the incipient iron curtain, the great body of Russians of the time were not merely prerevolutionary but prewar. It was not absurd to hope that a return to some sort of equilibrium with the West might still take place, in a manner similar to the return to normal life in France after the Terror and the Napoleonic Wars.

In one way or another, my taste for European contacts had grown with its own satisfaction. I was eagerly awaiting my next opportunity to see something more of the mother continent of our civilization.

3

1920–1925.

Years

of

Consolidation

When I came back from Strasbourg, I found my work well cut out for me. The Brownian motion papers were still in what we call a heuristic stage, which is to say that the general lines of organization of the subject and of the proofs of my theorems were already clear but that much work remained to be done before they could be considered complete. I showed my results to Professor E. B. Wilson, of the Massachusetts Institute of Technology, and at his advice I sent them at once to *The Proceedings of the National Academy of Sciences.*

Professor Wilson, who is now retired from teaching but still active in scientific administration, had been a pupil of Gibbs at Yale and had taught mathematics for some years at M.I.T. There he was teaching physics in 1920, and ultimately he came to be the mathematical specialist at the Harvard School of

Public Health. He has always been alert for what was new in the exact sciences, and for many years my enthusiasms have received his staunch support.

Another direction from which I received much encouragement was that of the department of electrical engineering, where Professor Dugald C. Jackson was head. I had known Professor Jackson and his son as neighbors when both of us had spent the summer of 1910 in New Hampshire. Jackson had been looking for some years for an engineer with mathematical sense, or for a mathematician with engineering sense, to resolve some of the problems impeding the theoretical electrical engineering of the time the state of which requires some explanation. Electrical engineering is divided into two more or less clearly separable fields, known in English as power engineering and communication engineering and in German by names which when translated into English become, respectively, "the technique of strong currents" and "the technique of weak currents."

Of these two fields of activity, power engineering had become fairly stabilized by 1920. Most of the types of electrical generators, motors, and transformers which now exist were already thoroughly understood by that date; and the present trend towards the fractional horsepower motor and the separately motored machine were well under way. What progress has been made in power engineering since 1920 belongs not to the tactics of the subject so much as to the strategy of large systems and electric supply. The great power nets in the United States and other countries have been increased, interconnected, and stabilized.

As to communication engineering, the situation was stabilized much later. Wireless had been an established art for twenty years, but for the most part this was wireless in the limited original sense in which Marconi had conceived it. Broadcasting was yet to come on a national scale, and the few preliminary attempts at radiotelephony were of more interest to curious young boy scientists and to hams in general than to the public. The electronic valve had indeed arrived, but there was scarcely a suspicion of the extent to which it was to modify our entire life. Television was not a new concept, for people had talked of television even before the beginning of the century, but it was just emerging from the stage in which it was conceived in terms of the selenium cell into that of practicable and rapid photoelectric apparatus.

The telephone, indeed, was triumphant everywhere, and was extending around the world the tentacles of a tight communication net. In the United States the A.T.&T., the master telephone company, was unrivaled throughout the whole scope of business in its financial magnitude and in its enlightened research policy. Thus, it was a natural time for a forward-looking electrical engineer like Jackson to devote a large part of his attention to problems in communication theory.

The logical basis of communication theory at that time was far from satisfying, and it appeared to be much less satisfying than it really was. It was of course understood that speech is carried on a telephone line by a fluctuating current whose fluctuations map those of the voice input. The great problem was to understand the full implications of the theory of fluctuating currents and voltages.

For several decades, the theory of fluctuating currents and voltages had dominated not only communication engineering but power engineering as well, in the form of the theory of the alternating current. The ordinary direct, continuous current is rather intractable. There are no simple means to step its voltage up or down, and when high-voltage direct-current lines have been used (as in France for instance), it has been necessary to run a number of generators in series against formidable problems of insulation and control.

The man who perhaps contributed more than anyone else to

a solution of the problems of generating and using alternating current was Nikola Tesla. This brilliant and eccentric Yugoslav engineer worked for the Westinghouse people and converted them to the policy of generating current not in a continuous stream but as a series of surges back and forth, say at the rate of sixty per second. With the aid of the transformer, this alternating current can be stepped up and down in voltage with the greatest ease, while it can be produced in generators free from many of the other annoying problems associated with direct current. It can be used in various types of motors, including certain sorts of induction motors entirely free from sliding electrical contacts. The only connection between the fixed winding of the motor, which is fed from the outside lines, and the moving winding, which is a part of the rotating apparatus, is the electromagnetic one which also exists between the windings of a transformer. As a matter of fact, in certain forms of the induction motor there is no electrical connection between the fixed winding and the winding of the rotor, or moving part. The electrical current which magnetizes the iron of the rotor is produced by the simultaneous action of the rotor and the fixed part, or stator, as an electromagnetic transformer. Such a piece of apparatus has the great advantage of not having any moving contacts whatever and is simpler, safer, and more troubleproof.

In the early days of the alternating current, there was a battle royal between the Westinghouse people, who owned the alternating current inventions, and the General Electric and Edison people, who had invested heavily in direct-current engineering. This quarrel had as one of its consequences the fact that New York State decided to execute criminals by alternating current. This was the result of a deal put through the legislature in order to give a bad name to the supposedly more dangerous alternating current and to make people unwilling to have this introduced into their houses. Before long, however, the quarrels between the two schools of electrical engineering were patched up, for alternating current became available to the General Electric interests as well as to the Westinghouse Company.

It was in fact at General Electric that the theory of alternating current and alternating-current networks became organized and consolidated by Charles P. Steinmetz. This brilliant little man made great use of the mathematical theory of imaginary or complex numbers (which are quite as genuine and actual as real numbers) to describe alternating currents and voltages and the apparatus operating on them. The reason for introducing complex numbers into the engineering of alternating currents is that each complex number really consists of a pair of real numbers—the so-called real and imaginary parts—while an alternating current of given frequency is also determined by two real numbers, one of which gives its intensity and the other the phase or time at which it passes through zero.

For many years the theory of alternating-current engineering has been pretty complete, at least as far as concerns currents and voltages of fixed frequency, such as sixty-cycle currents and voltages. In telephone and other communication engineering we also deal with a sort of alternating current, but this alternating current is far more complicated because its frequency in oscillations per second is not fixed and because at any given time we must deal with many simultaneous sorts of oscillation. A telephone line carries at the same time frequencies of something like twenty per second and frequencies of three thousand. It is precisely this variability and multiplicity of frequency which makes the telephone line an effective vehicle of information. The line must be able to carry everything from a groan to a squeak.

Here we are concerned with one of the most ancient branches of mathematics, the theory of the vibrating string, which has its roots in the ideas of the Greek mathematician Pythagoras. He and his disciples knew very well that the vibrations of strings produce sounds and that there is a connection between the pitch of the sound generated, the length of the string, and its density and tension. To what extent the Greeks were aware that a single string can vibrate in several modes at the same time I do not know; but this fact was thoroughly familiar to the early modern scholars of the seventeenth and eighteenth centuries.

The fundamental notion in all this is what we call the sinusoid, and to explain this let us suppose that we have a drum of smoked paper turning around and let us further suppose that we have a tuning fork vibrating parallel to the axis of the drum, and that to the end of this tuning fork is attached a straw which will make a white mark on the smoked paper. As the drum revolves at constant speed, the straw will leave an extended mark which we call a sinusoid.

Let us now consider more complicated curves, made up by adding sinusoids. It is possible to add curves to one another by adding their displacements, that is to say, by combining two tuning forks of different rates of oscillation so that they both act on the same straw as it traces its path along a drum of smoked paper. In this motion we can observe two or more rates of oscillation in the same curve at the same time. The study of how to break up various sorts of curves into such sums of sinusoids is called harmonic analysis.

There is a fundamental theorem that says that if we have a curve which repeats the same form indefinitely it can be broken up into an infinite number of separate sinusoids which repeat themselves at different rates. While results of this sort were known in the eighteenth century, the name generally connected with this theorem is that of Fourier, a member of the French Academy of Sciences who accompanied Napoleon on his expedition to Egypt.

Fourier's name is also connected with other ways of adding together sinusoids in which the number of sinusoids to be added is too great to be represented by a first curve, a second curve, a third curve and so on. We may indeed have to add a mass of sinusoids which is entirely too dense to be arranged in one-two-three order.

Two parts of harmonic analysis concern themselves respectively with the analysis of periodic processes, which is given by what is known as Fourier series, and the analysis of processes which come up from zero in the course of time and which go down to zero again. In both cases the mathematician is forced to use the sophisticated methods of adding quantities to which we have already referred under the name of Lebesgue integration.

The really satisfactory theories of the Fourier series and integral were too new in 1920 to have trickled down to the working electrical engineer. Moreover, the sort of phenomenon in which the engineer is chiefly interested had almost entirely escaped the treatment of the pure mathematicians. The Fourier series, which the pure mathematicians had treated, was useful only for the study of those phenomena which repeat themselves after a fixed time. The standard form of the theory of the Fourier integral, as developed by Plancherel and others, concerns curves which are small in the remote past and are destined to become small in the remote future. In other words, the standard theory of the Fourier integral deals with phenomena which in some sense or other both begin and end, and do not keep running indefinitely at about the same scale. The sort of continuing phenomenon that we find in a noise or a beam of light had been completely neglected by the professional mathematician, and had been left to such mathematically-minded physicists as Sir Arthur Schuster of Manchester.

I came to understand that the various demands made on me by Professor Jackson concerning the proper foundation of communication theory were to be fulfilled by a further study of harmonic analysis but that this could not be done solely on the basis of harmonic analysis as it existed at the time. What the communication engineers actually did was to use a formal calculus of communication theory which had been developed some twenty years before by Oliver Heaviside. This Heaviside calculus had not as yet been given a thoroughly rigorous justification, but it had worked for Heaviside and for those of his followers who had absorbed the spirit of his theory sufficiently to use it intelligently.

For several years the chief demand made on me at M.I.T. by the electrical-engineering department was to put the Heaviside calculus on a proper logical foundation. Other people were doing the same thing at the same time in other countries, although I do not think that any of these treatments were more satisfactory than the one which I ultimately gave. In performing this task, I had to study harmonic analysis on an extremely general basis, and I found out that Heaviside's work could be translated word for word into the language of this generalized harmonic analysis.

In all this there was an interplay between what I was doing on the Heaviside theory and what I had done on the Brownian motion. Previous to my work there had been no thoroughly satisfactory example given of the sort of motion that would correspond to sound or light with a continuous spectrum—that is, with energy distributed continuously in frequency instead of being lumped in isolated spectrum lines. The harmonic analysis which had already been given corresponded more closely to what one sees when one examines the light of sodium vapor than what one sees when one examines sunlight. (The light of sodium vapor is concentrated in a number of bright lines, whereas sunlight has a continuous distribution in color and consequently in frequency.)

In Chapter 1 I pointed out my work on the mathematics and physics of the discrete, and in particular of the Brownian motion, in which we study the succession of impacts which a particle in a gas receives from the moving molecules—the shot effect, which is due to the manner in which electric currents are conveyed by a stream of individual electrons. I found that it was possible to generate continuous spectra by means of the Brownian motion or the shot effect and that if a shot-effect generator were allowed to feed into a circuit that could vibrate, the output would be of that continuous character. In other words, I already began to detect a statistical element in the theory of the continuous spectrum and, through that, in communication theory. Now, almost thirty years later, communication theory is thoroughly statistical, and this can be traced directly back to my work of that time.

My interest in harmonic analysis did not exhaust my mathematical activity. There were other problems which occupied me, some intensively and some in a more or less desultory way. The combined-research group of our department had begun to achieve a bulk of publication which made a journal of our own desirable, and we had already embarked upon this project.[1] I was the first acting editor, but the responsibility was soon taken over by Philip Franklin, who had recently come to us from Harvard, and who had been my friend and associate of Aberdeen Proving Ground days.

I used to talk occasionally with Professor O. D. Kellogg of Harvard concerning problems of possible interest on which I might do research. I did not realize at that time how carefully many professors conserve problems for their own graduate students and how sharply they regard proprietary rights in new problems. I had been used to the freer atmosphere of England and to the lavish manner in which my father had scattered the seeds of his ideas before all who would listen. My active and assertive curiosity did me no good in the opinions of those who might have benefited me by their esteem. I was not officially

[1] I was lucky to have a journal at my disposal in which I could secure quick publication.

a student of Kellogg's. He was very helpful to me, but I consumed too much of his time, and I think that he considered me as something of a nuisance.

I learned from Kellogg that the old problem of potential distribution was attracting renewed interest. It would be of no point to state the problem here explicitly, but it is quite possible to tell the layman what sort of a problem it is. There are many physical questions which involve measurable quantities that vary over a plane or over space. The temperature in a room is such a quantity, and there are certain other similar quantities related to the flow of a liquid or the diffusion of a gas, which I can measure with a voltmeter, which gives the various electromotive forces between points in a room and the ground or between one point and another in a conductor in which an electric current is flowing.

Here I can dispense with a complete definition of electromotive force; for it is necessary to know only that electromotive force is what we measure in volts. The mathematics of all quantities varying over space and time lies in the field of partial differential equations, which is a mathematician's way of saying that we have various relations among the rates at which these quantities fall off in different directions and the rates at which they change in time. Ever since the time of Leibnitz it has been well known that there are quantities distributed both in time and in space; and that they have space rates of change as well as time rates of change. The temperature may change at so many degrees per hour; but it may also change at so many degrees per hundred miles we go north or at so many degrees per hundred miles we go east. Again, if water is flowing downhill, the steeper the hill, the faster the water flows.

Many quantities thus distributed in space and in time are of great engineering importance. It is the rate at which the local electromotive force falls off as we leave a transmission line which determines whether the line is going to conduct electricity without substantial leaks or will glow at night with a corona effect which represents many dollars out of the pocket of the electric company and its subscribers. The study of the heat-insulating powers of a house wall depends on relations among the flow of heat, the rate of fall-off of temperature, and so on.

Much of the mathematics of these quantities (which are known as potentials) is clear and direct. In that part of a room which is away from the walls, or from any other conductors, the problem of the distribution of electromotive force is relatively simple. However, when we come to the immediate neighborhood of regions of the room with very special electric properties, we get into trouble. Near these regions, which are known as boundaries, the problem of electrostatic potential reaches a new order of complication. Similar difficulties arise in the related theories of temperature and of fluid flow.

In the case of the electrostatic potential, one particular boundary phenomenon is exhibited by the sharp-pointed conductor, such as the lightning rod. Around such a sharply-pointed conductor projecting into a region in which there are electric charges, the rate at which the electromotive force drops off becomes enormous or even infinite. The electric field will not hold such rates of dropping off of potential, or potential gradients as they are called. Around such a sharp point the air is continually breaking down as an insulator, and if the field is large, a distinct corona effect will be seen in the dark. Many sailors have observed the curious effect known as the corposant, where nails and other pointed objects glow with a ghastly light in the electrified atmosphere of a thunderstorm. It is through something like this corona effect that a lightning rod relaxes the potential gradient of the charged atmosphere about it, by a gradual and unspectacular process, before the tensions build up to such a point that they may cause a disastrous stroke.

At places at which the voltage changes rapidly in space,

certain media will be strained and will break down as the air does when a stroke of lightning passes or as a piece of glass when a stroke of lightning bursts through a window. The ability to stand up to these stresses is known as the dielectric strength.

So far I have been stating the problem of the pointed conductor in a physical way, depending on the specific dielectric strengths of the different media into which the conductor may be pointed. There are however closely related problems of a more formal and purely mathematical character.

We here come to one of those mathematical situations in which there is a close relation between mathematical and physical ideas but in which the correspondence between the two is not precise. All physically pointed objects are like the point of a needle that is very slightly rounded off at the end. It is possible, however, to imagine a still sharper point—such as one we might obtain by revolving about its middle line the cross section of an infinitely sharp hollow-ground razor—in which the two profiles are tangent to each other. Impossible as it is to realize such a figure physically, its conception offers no mathematical difficulty whatsoever. It is even possible to consider an electrical potential distributed about such a re-entrant spike and to ask how such a potential would behave near the very point of the spike.

It will be found that there are cases where the mathematical behavior of potential around this ideal spike will be strongly suggested by the actual behavior of potential about a very sharp physical spike. In the physical case the strains become so great that the matter in the field breaks down. In the mathematical case there need be no matter in the field to break down, but the field itself may become discontinuous. If this is so, the potential at the point of the spike becomes indeterminate and assumes one value if we reach the point of the spike by one path and another if we reach it by another path. It was this phenomenon which I started to investigate at Kellogg's suggestion. My object was to determine for what spikes this discontinuity could occur.

A Polish mathematician by the name of Zaremba had obtained here certain results which gave some hypotheses of sharpness that were enough to ensure the indeterminacy of the potential, and some other hypotheses of bluntness which were enough to ensure that the potential would not become indeterminate; but these conditions left between them a gap in which our knowledge was incomplete. In this intermediate field Professor Kellogg had done vitally important work, and two of his young friends were writing doctoral dissertations on the subject at Princeton. When Kellogg informed me of the work that was being done in this corner of potential theory, I immediately began to reflect how I myself should attack the problem.

I very soon found that I could make quick headway in this subject and that in fact, within a matter of days, I had got even further than the two Princeton doctoral candidates. When I showed Professor Kellogg my mathematical results, there was a sudden change in his attitude. At the beginning I had found him rather pleased with my interest in potential theory. Now, however, he was chiefly concerned with the effect that my pursuit of the subject might have on the acceptance of the doctoral dissertations of the other two young men.

There is an understanding at many schools which was at that time a general convention—that the publication of a doctoral dissertation is an indispensable condition of its acceptance. It is clearly more difficult to publish a paper which has been anticipated than to publish one when the results are known to be completely new. This I felt to be unjust, and I considered that the only sane criterion of originality from the standpoint of a doctor's degree should be whether the paper was new as compared with the literature reasonably available at the time of its submission. Here, by "reasonably available," I mean reasonably available in view of the actual opportunities of the author.

I am afraid I did not accept with any grace Professor Kellogg's dictum that I was to erase from my mind the work I had already done on the potential problem and to clear the tracks for the two doctoral candidates. I had been aware through Dr. Kellogg's leak, and only through Dr. Kellogg's leak, that other people were working on the problem, but I had no information as to their methods and tools, and my result was genuinely original.

Furthermore, I did not accept with alacrity Kellogg's suggestion that I was now an established mathematician, who did not need these papers and who ought to give them up as a charity to youth and inexperience. Both of the candidates were older than myself, and both of them enjoyed the secure position of being the pupils of men with influence in American mathematics. I had never received favors from those in power, and it was only when it was to my disadvantage that any Harvard man thought of treating me as an established mathematician.

If I had had my full, undivided attention to devote to my scientific work and my scientific position, the situation would have been difficult enough. However, the scientist is at the same time a human being, and his human needs will not wait indefinitely on his scientific career. I was by now in my late twenties and I had begun to look forward to the fuller life of a married man. I was already beginning to pay attention to the young lady who is now my wife.

This young lady who so interested and attracted me was Margaret Engemann. She came from a family in Germany that was close to the land, and which had been gradually emerging from the status of small farmers to that of renters and stewards of great estates, to the clergy, and to professional life in general. Her mother had come over to America after the death of her husband and had lived an active, romantic, outdoor life in the new West. It was the directness, genuineness, and sincerity of the mother that I saw reflected in the daughter, and which gave to me first the promise and then the certainty that this was the girl for me.

I had visited her and her family at about the time of my run-in with Kellogg on a raw and bleak December day. In waiting for the streetcar home, I had got thoroughly soaked and had developed the beginnings of a bad cold. When I talked over the problem of the publication of my paper with Kellogg when we met at a local meeting of the American Mathematical Society, I was already sick and half-delirious with what later turned out to be pneumonia. Instead of acquiescing in his point of view, I felt that it was an unfair assertion of solidarity among the insiders against the outsider and asserted my intention of securing quick publication for my results in our new mathematical journal. This started a storm of antagonism against me, and both Kellogg and Birkhoff thundered at me from an exalted moral elevation.

I felt thoroughly sick and discredited. The next day I commenced a chilly weekend of winter sports at the Groton farm which my parents had bought for their ultimate retirement. I returned home and went immediately to bed, where I found that I was down with a first-rate case of bronchopneumonia. All through the pneumonia, my delirium assumed the form of a peculiar mixture of depression and worry concerning my row with the Harvard mathematicians and of an anxiety about the logical status of my mathematical work. It was impossible for me to distinguish among my pain and difficulty in breathing, the flapping of the window curtain, and certain as yet unresolved parts of the potential problem on which I was working.

I cannot say merely that the pain revealed itself as a mathematical tension, or that the mathematical tension symbolized itself as a pain: for the two were united too closely to make such a separation significant. However, when I reflected on this matter later, I became aware of the possibility that almost any experience may act as a temporary symbol for a mathematical situation which has not yet been organized and cleared up. I also came to see more definitely than I had before that one of the chief motives driving me to mathematics was the discomfort or even the pain of an unresolved mathematical discord. I became more and more conscious of the need to reduce such a discord to semipermanent and recognizable terms before I could release it and pass on to something else.

Indeed, if there is any one quality which marks the competent mathematician more than any other, I think it is the power to operate with temporary emotional symbols and to organize out of them a semipermanent, recallable language. If one is not able to do this, one is likely to find that his ideas evaporate from the sheer difficulty of preserving them in an as yet unformulated shape.

It was at the period of my illness that I became really aware of how much I needed the young lady who was later to become my wife. I will not say that from this time on my courtship followed a direct path, or that I became immediately sure that I wanted marriage; but at any rate I had embarked on the series of emotional changes and ups and downs which ultimately lead to our marriage. I have told the story of this in my earlier book, *Ex-Prodigy*. I shall return to it here only insofar as it is directly relevant to the incidents of my career as a scientist.

It is as a scientist that I am exhibiting myself to the public in the present book, and I shall deliberately play down even those emotional events of my own life and of our later life together which have been most important to us. I wish no reader to draw the conclusion that my emotional life has been restricted to my scientific career, or that I could live with any satisfaction without the loyalty, affection, and continued support of my lifelong companion. What is inhibiting me here is the intensely personal quality of the enduring love and companionship between man and wife. I cannot convey this to my readers without what I feel to be an infringement of what belongs to us two alone and of what, because of its very reality, depth, and lasting endurance, is too free from casual vicissitudes to contain much that we care to impart to anyone but ourselves. My science belongs to the world, my home life and affections to no one but to those dear to me, and to myself.

Let me return to the general question of competitiveness among mathematicians and of the ethics which governs it. I have already said that I was more or less repelled by the high pressure of work which was bound to be competitive from the start, yet I knew very well that I was competitive beyond the run of younger mathematicians, and I knew equally that this was not a very pretty attitude. However, it was not an attitude which I was free to assume or to reject. I was quite aware that I was an out among ins and that I would get no shred of recognition that I did not force. If I was not to be welcomed, well then, let me be too dangerous to be ignored.

I was not alone in my competitiveness. At least one of the greatest American mathematicians, a man whose disapproval was the highest hurdle I should have to leap, was even more intensely competitive than myself. I have always taken in ideas easily and given them out freely, and, although emulation has been a part of my nature, I have not sought to work in the profoundest secrecy and to spring my new results on a world which has not even known that I have been working on them. In this I offer a strong contrast to certain older colleagues, who may have shown less naïve joy in the immediate results of their scientific conquests, and I have never tried to steer other investigators away from my own work so that I could be the beneficiary of the surprise effect of a new paper carefully guarded until I could present it with a maximum impact. I have not been more competitive than many of those about me, but I have been less meticulous in presenting the appearance of lacking competitiveness and I have not been careful to mend my fences.

4

The Period of
My Travels Abroad—
Max Born
and
Quantum Theory

It was only two years after my trip to Strasbourg that I began
to resume my visits to Europe. Part of my eagerness was due
to my desire to drink deep at the fountain of European mathematics, of which I had already tasted, but another part came
from family considerations.

Soon after the end of the war my parents had bought a farm
house and apple orchard in the town of Groton, Massachusetts,
with the hope that they might live there after my father's retirement. This farm was indeed at the disposal of the entire
family for our recreation; but it would have taken the full
efforts of the family to run it, and even in our vacations we were
no longer a group of unoccupied youngsters. We needed our
free time to recuperate from the very hard work of making our
way in our professional careers. It was scarcely reasonable to

expect that we should turn our well-earned holidays over to the job of raising vegetables and clearing our wood lot.

My sister Constance was teaching mathematics at Smith during much of the time, and my sister Bertha was studying chemistry at the Massachusetts Institute of Technology. Constance, a small, vigorous, self-confident young woman, was the mainstay of my parents and was supposed to be the one of us with the most *savoir faire*. In those matters in which my point of view came to differ more and more from that of my mother, Constance leaned heavily to my mother's side.

Bertha, seven years younger than I, was perhaps the most independent member of the family. She had been too young to undergo the full educational pressure to which I had been subject or even the more diluted form of it which Constance had experienced. During her secondary-school career, moreover, the education of my younger brother Fritz had been the chief family problem; and Bertha more than the other two of us had been left to her own devices. Thus Bertha has always been able to see the family problems from a somewhat more objective point of view than either Constance or I myself, in my early days at any rate.

I was eager to pass on to my sisters some of the pleasures of my European travels, and they were only too pleased to share these with me. I will not go into the chronology of the successive voyages which I made alone and with my sisters, but the summers of 1922, 1924, and 1925 found me abroad again, looking up friends of the family and various colleagues. I saw more and more of Lévy on these trips and I developed important new contacts not only in England and France but in Germany as well. I witnessed the great inflation in Germany during the summer of 1922 and found it a really terrifying experience.

My work on potential theory continued in two directions. First, I arrived at a new concept of the relations between the electromagnetic potential in the interior of a region and that on the boundaries. As I have already suggested, the earlier concept of the electromagnetic potential in the interior of a region was that it should completely fit the values given on the boundaries. However, I followed out notions akin to those of generalized integration, which I had already studied, and noticed that the interior potential of a region could be regarded as determined by an additive combination of potentials around the boundaries, even when this definition might fail to yield a continuous function as we approach the boundary point. This was a radically new idea, and it led me to significant extensions of many of the notions of potential theory, including those of charge and capacity. In this work I was motivated by concepts which belonged to the generalized theory of integration of P. J. Daniell, of which I have already written. The novelty was that I conceived the relation of the potential of an interior point to the boundary values as a sort of generalized integration rather than as a limiting process by which the internal potentials should be united with those at the boundary point. This was something of an inversion of the existing point of view on boundary problems. Like so many inversions of points of view in mathematics, the reformulation of the problems of potential theory let a breath of fresh air into a situation which had been deadened for years by too conventional a statement.

My old friend and mentor, Professor H. B. Phillips of M.I.T., had already studied quantities analogous to potential on square nets like screens and on cubical structures representing generalizations of these square nets. With the aid of my new generalized concepts in potential theory, I was able to make Professor Phillips's work an important step towards a universal theory of potential.

In this manner I added a considerable number of new and sharply definable concepts to the armament of potential theory. When I applied these to the old problem of Zaremba, which still had not reached a final solution, I found that they fitted.

This was at about the time at which the *Comptes Rendus* of the French Academy of Sciences began to be filled with papers on the theme of the Zaremba theory. These were written by Lebesgue himself and by a young pupil of his, G. Bouligand.

In many scientific subjects there comes a time when the sharpness and definiteness of the new papers indicate that an important goal is about to be achieved. So it was with the work of Lebesgue and Bouligand. I knew that if I did not put forth a maximum effort, the whole topic was soon to be crossed off the account books of mathematics as one finally resolved. Accordingly, I put forth a maximum effort, employing those new tools which I had made my own, and I was delighted to find that I had achieved what was from the standpoint of research at that time a complete solution of the problem.

I was aware that I must hurry. I went to a Mexican student of mine, Manuel Sandoval Vallarta, who was much more proficient in French than I was, and got him to help me to rewrite my ideas in acceptable French. I mailed a brief note to Lebesgue for publication in the Comptes Rendus and awaited the outcome.

What followed is a coincidence of a sort much commoner in the history of discovery and invention than one might suppose. While my letter was crossing the ocean, Bouligand obtained some extremely important results which he had not yet had time to polish up. He took council with Lebesgue, who advised him to submit these results to the academy in a sealed envelope, after a custom sanctioned by centuries of academy tradition. The very next day, my paper came in and Bouligand's sealed envelope was opened. The two papers appeared side by side in the next number of the Comptes Rendus, with a preface by Lebesgue covering both of them. While they were expressed in different mathematical language, the main idea of the two was identical. However, the logic of Bouligand's paper was not as complete as mine, owing to the fact that what he had sent in was only a preliminary communication for purposes of record, and not a polished and finished job.

This was even more of a dead heat than my previous double discovery of Banach space. Both the Banach space competition and that with Bouligand turned out to be extremely friendly. Bouligand was even more ready than I was to admit the somewhat greater completeness of my paper, and we made arrangements to meet when I should come to France on a visit.

A piece of work which started out under less competitive pressure but which also ended in a friendly way was stimulated by the research of the Danish mathematician Harald Bohr on what he called "almost periodic functions." These represent curves which do not quite repeat themselves after the fashion of a pattern on wall paper but come close to doing so. Their discovery represented an important extension of harmonic analysis. I myself, as I have said, was working on extensions of harmonic analysis to which I had been led through an attempt to justify Heaviside's formal calculus. Again there was nothing for me to do but put on steam and see if I could consolidate my ideas in the field. I did succeed in doing so, and I obtained an adequate theory which covered not only those spectra of light which can be represented as sharp lines, such as we find in parts of the spectra of the elements, but also those where the power is distributed continuously. Bohr's theory was valid only for line spectra. With the aid of one or two tricks, from my own theory I was able to deduce not only the Bohr theory but a much wider range of results concerning continuous spectra as well.

In this I had to use ideas very closely akin to those which I had used in the study of the Brownian motion. In particular I had to make use of curves which are continuous but which are so crinkly that they can not properly be said to have a direction. I have already pointed out in my discussion of the Brownian motion that these curves had been more or less the step-

children of mathematics and had been regarded as rather un-
natural museum pieces, derived by the mathematician from
abstract considerations, and with no true representation in
physics. Here I found myself establishing an essentially physical theory in which such curves played an indispensable role.

There was another contact with European mathematics
which I made at this time in quite an unanticipated way,
though in a personal rather than a scientific matter. I had seen
a good deal of the work of Leon Lichtenstein, a German mathematician who had been studying the theory of fluid flow, and
who was the editor of what was then the chief international
review journal for mathematicians. Father had known of a
cousin Leon who had gone to the same institute of technology
at Berlin at which he himself had studied, and who had found
the atmosphere of industrial research not too agreeable. Father
was also aware that Leon had left industry for academic work
in applied mathematics but was unaware either of his degree
of success or where he was now working.

We received a letter from an aunt of mine in New York,
telling us that cousin Leon had turned out to be rather more
successful in mathematics than we had anticipated. The letter
gave us his full name, which was Leon Lichtenstein. I put two
and two together and thought it highly likely that cousin Leon
and the famous mathematician were the same person. I wrote
to Lichtenstein, asking him directly if he was our cousin, and
I received a very friendly reply. He was indeed the Lichtenstein of whom Aunt Charlotte had written, and he was quite
aware of my existence and my work. He invited me to look
him up the next time I should find myself in Europe. He still
lived in Berlin, but his teaching was at the University of Leipzig. Here, as I afterwards learned, he was dean of the Faculty
of Sciences.

The summer after I had made my contacts with Bouligand
and Lichtenstein I was in Europe with my sister Bertha. I ran

down to Poitiers to see Bouligand. He was waiting for me at the station with a copy of one of my articles held up so that I could identify him. He proved to be an unpretentious young Breton, and he took me into his family for the visit. There was much to see in Poitiers, which is a charming, romantic town with most interesting architectural monuments. Bouligand introduced me to a friend of his who was professor at the *lyceé* and an authority on local antiquities, and between them they showed me all the tidbits of interest.

I also went to Germany that summer to see cousin Leon. Like Bouligand, he had to identify himself to me, and I had to identify myself to him, because we never had even seen pictures of each other. He met me at the station, holding up a sheet of paper on which he had written, in my honor, the chief formula in potential theory.

Leon Lichtenstein was bald and bearded and did not have much facial resemblance to my father, but like him he was a short, vigorous man, with brisk motions and strong opinions. He was very anti-American in many of his attitudes, although cordial to me personally. Bertha and Mrs. Lichtenstein, who was the balance wheel of the Lichtenstein family, had considerable difficulty in keeping the arguments between Leon and me from degenerating into quarrels.

Here I ran into one of the minor problems of life in Germany. At our first meeting, Leon had asked me to use *du*, the familiar form of German address, with him. Mrs. Lichtenstein, on the other hand, had made no such suggestion to me, although she was quite as cordial as her husband. Under these circumstances I did not feel that I was entitled to the same degree of familiarity with her as with my cousin, and I stuck to *Sie*, the formal mode of address.

I visited Göttingen in 1924 for old times' sake and found that my new ideas had begun to interest the Göttingers. In 1925, after a little mountain climbing with Alexander, of Princeton,

I again visited Göttingen, where my work on generalized harmonic analysis was beginning to attract real attention.

The new head of all Göttingen mathematics was Richard Courant, an industrious, active little man, who was eager to keep all the strings of mathematical administration in his hands. Courant suggested to me that I might find some American sources of funds for a year's study in Göttingen, and for research under continued contact with my Göttingen colleagues. The John Simon Guggenheim Foundation had just been established in New York, and Courant pointed out to me that that was an appropriate benefactor to which I might turn for funds. He promised me the full co-operation of my Göttingen colleagues in making my trip agreeable and in providing me with an assistant to help organize my papers and to take care of my lapses in German.

He sent me over to pay my respects to Felix Klein, who shared with Hilbert the prestige of being one of the two leading Göttingen mathematicians. Klein was in poor health and, as a matter of fact, had only a few months of life ahead of him. I welcomed this opportunity to establish yet another connection with the august past of mathematics.

I began my visit to Klein by a social blunder. When the elderly housekeeper appeared at the door, I asked in my best German, *"Ist der Herr Professor zu Hause?"* She replied in a tone of rebuke, *"Der Herr* Geheimrat *ist zu Hause,"* thus reproaching me for addressing a privy councilor by the lower title of professor. In German science, I may say, the social position of a Geheimrat was like that of a scientist in England who had been knighted; but I have never seen so much fuss made about a knighthood in England, as was made in Germany about the title of Geheimrat.

I went upstairs and found Felix Klein in his great study, a pleasant, high, airy room lined with bookcases and with a large table in the middle covered with an orderly disorder of books and open periodicals. The great man sat in an arm chair behind the table, with a rug about his knees. He was bearded, had a fine, chiseled face, and carried about him an aura of the wisdom of the ages. He spoke with a noble condescension, as if he were a king; and as he spoke the great names of the past ceased to be the mere shadowy authors of papers and became real human beings. There was a timelessness about him which became a man to whom time no longer had a meaning. After I had listened respectfully for a few minutes, I found that I was given my *congé* as I might have been given it at a court.

The talk which I gave the Göttingen people on my work on general harmonic analysis was very well received. Hilbert, in particular, showed great interest in the subject, but what I did not realize at that time was that my talk was closely keyed to the new ideas of physics which were about to burst into bloom at Göttingen in the form of what is now known as quantum mechanics.[1]

Quantum mechanics was a subject in mathematical physics which had originated in 1900 in the work of Max Planck on the equilibrium of radiation in a cavity. In plain language, the subject matter of quantum theory is the study of such light as we find inside of a hot furnace after light and hot matter have come to equilibrium so that if we look into a cavity with heated walls, such as a blast furnace, the light coming from inside the furnace changes in character as the temperature changes. This is a readily observable effect which we all know from the difference between a red-hot piece of metal and a white-hot piece of metal. The spectrum of the light coming from the red-hot furnace ceases somewhere in the red or yellow, but the light

[1] In this chapter I must face the problem of trying to explain in nontechnical language what the technical meaning of some very difficult work has been. The reader who is not interested in a detailed documentation of my work at this time is advised to skip over the technical passages.

coming from the white-hot furnace may go far into the ultra-violet.

The nub of the difficulty in explaining this relation between light and heat, which Planck solved by a brutal new hypothesis, was that the traditional representations of light as a continuous phenomenon was not satisfactory. In light as in matter, he argued, there is a granular rather than continuous texture.

The earlier physics had not been able to conceive any mechanism by which the color distribution of light in a furnace could be determined by the furnace's temperature. Planck's eventual explanation of this easily observable phenomenon was, however, not simple. It was associated with ideas concerning mathematics and thought in general which go back to the end of the seventeenth century, during a period when an important intellectual battle was being fought between the atomists, who believed in the discreteness of matter, and those who believed that matter is continuous. There were various philosophical considerations which made this debate especially critical.

It was not, however, so much the general philosophical climate of the time as a technical innovation which brought the dispute to a head. This innovation was the discovery of the microscope by the Dutchman Leeuwenhoek, who had perfected his device to the point at which he could see something of the teeming life in a drop of stagnant water.

The discovery of a new instrument often leads immediately to a new insight. Before Leeuwenhoek, the study of living organisms had been limited to what could be seen by the naked eye or, at best, with a primitive hand lens. Thus, scientists, while they might have had the Democritean idea that the world exists of extremely small particles or atoms, had made no considerable progress in seeing phenomena smaller than, say, a grain of sand.

Leeuwenhoek's microscope showed by direct observation

that a drop of pond water was a teeming mass of life suggestive of a crowded city. The new power lent to the eye engendered a new range of imagination, and everyone's thoughts turned to the fine structure of the world and to the philosophical implications suggested by the process of magnification. One result of this experience, perhaps, was Swift's famous jingle:

So, naturalists observe, a flea
Hath smaller fleas that on him prey;
And these have smaller still to bite 'em;
And so proceed *ad infinitum*.

The background of this little jingle is more interesting than the jingle itself may seem to us at this late date, for among the objects that Leeuwenhoek studied with his new microscope were the spermatozoa of man and the animals, which Leeuwenhoek quite reasonably interpreted as playing a part in conception. Through the imperfect microscopes of Leeuwenhoek and his followers, however, it was easy to imagine that the spermatozoon contained a small, rolled-up fetus. This theory gave a plausible interpretation to the act of conception, for it was believed conception consisted merely in the implantation of the spermatozoon in the womb, in which environment it could grow in size till it became an embryo of the sort which was already familiar to the doctors. The idea that the spermatozoon was the sole antecedent of the embryo led to some very interesting biological speculations.

If the spermatozoon was itself an early stage of the fetus, it was natural to think that it was a human being in miniature, with all the organs of the human being on a smaller scale, distorted but still essentially there. By this token, it should contain smaller spermatozoa, much as Swift's flea carried lesser fleas on a scale far smaller than the microscope of the day could show. These in turn could be thought to contain still smaller spermatozoa, and so on ad infinitum, so that the whole future of the human race actually lies preformed within the bodies of those now existing. This preformationism argued for an infinite subdivisibility of matter, and the philosophical consequences of this were eagerly studied, particularly by the great philosopher, Leibniz.[2]

[2] The step from Leibniz to Swift involves certain aspects of the history of the early eighteenth century which deserve comment. Leibniz was a great philosopher and physicist by avocation, but his official position had been that of archivist to the court of Hanover. In this position he showed himself to be not only a librarian but a diplomat of the first rank, eager for the welfare and the aggrandizement of his ruler. There is much to be said for the conjecture that he was active in the negotiations which put the house of Hanover on the throne of England. Since it was the Whig party in England which desired the coming of the Hanoverians, in order to terminate the reign of the unpopular Stuarts, Leibniz became identified with the Whig intrigues. His contact with England was greatly facilitated by his membership and his active share in the Royal Society.

Swift, on the other hand, was a Tory supporter of the Stuarts, and he took an active share in the attempted *coup d'état* which tried to put the Old Pretender on the throne after the death of Queen Anne. Thus, Leibniz and Swift were key figures, respectively, of the two conflicting parties in the English politics of the day. It is no wonder that a great antagonism grew up between them.

This antagonism is shown in the third of the four books of *Gulliver's Travels*, the voyage to Laputa. Many people have wondered at the virulence with which Swift lashes scientists, these impractical projectors who measure a man with a sextant to fit him with a suit of clothes, who extract sunbeams from cucumbers, and who attempt to attain all the learning of the ages by a process equivalent to Eddington's monkeys and typewriters. In fact, they represent nothing but the Royal Society and in particular the Leibnizian influence in the Royal Society. It is thus not astonishing that one of the targets of Swift's wit should be the essentially Leibnizian situation of the fleas upon the fleas and so on ad infinitum.

This was not the only place at which Swift showed himself fasci-

Leibniz conceived the world after the analogy of the drop of water and the similarly teeming drop of blood as a plenum. That is, he conceived that all the apparent spaces between living beings and within living beings are themselves filled with living beings on a smaller scale. This theory led Leibniz to postulate the infinite subdivisibility of life and, consequently, the continuity of matter.

This opinion, which was generated as we have seen by the microscopic observations of his day as well as by the inner workings of his own philosophy, led Leibniz eventually to a new interpretation of mathematics. He was, we must remember, one of the co-inventors of the calculus, and he originated the notation which we use even now. For him not only are time and space infinitely subdivisible, but quantities distributed in time and space may have rates of change in all their dimensions. For example, one quantity distributed in time and space is temperature. When I say that the thermometer is dropping at the rate of ten degrees an hour, I am speaking of its time rate of change. When I say that it is dropping at the rate of three degrees per mile as I go west, I am giving one of its space

---

nated with the problem of the variable scale of nature and of what would happen to the world and the individuals in it under a contraction or an expansion. It is likewise the theme of the first two books of *Gulliver's Travels*, the voyage to Lilliput, where the inhabitants are one twelfth the height of a normal man, and the voyage to Brobdingnag, where the inhabitants are giants seventy feet tall.

In both cases, Swift's imagination concerning the effect of change of size is keen but limited. It applies to the physical dimensions but not to their powers of motion. He is not aware that the Lilliputians, if they were made of human flesh and blood, should be able to jump a height several times greater than their own, nor of the similar fact that his Brobdingnagians would be so slothful and earth-bound that they would be scarcely able to stand up.

rates of change. In discussing quantities which have a distribution both in time and in space, a natural mathematical law is the partial differential equation in which time rates of change and space rates of change are related to one another in a system where time and space are both infinitely subdivisible. Thus, Leibniz, in arguing for the continuity of the physical world, became the spokesman for a view in direct contradiction to atomism.

The development of physics since his time has brought both atomism and the continuistic theory to a perfection and to a sharpness of opposition which they did not possess in his day. The molecule has been all but seen, and the chemical evidence for the existence of the discrete atom is clear. Beyond the atom, new vistas of atomicity have been discovered in the electron, the proton, and the many new fundamental particles discovered in the atomic nucleus; while in the meantime the continuum theory has become a useful and almost indispensable tool for the study of the dynamics of gases, liquids, and solids and for the theory of light and electricity. That these two great directions of thought have come into head-on collision with one another has led to some of the chief problems of modern physics.

This collision began to take shape about a hundred years ago, when Clerk Maxwell developed what is now known as the kinetic theory of gases. A gas consists of particles called molecules which can move in several independent ways. A molecule can move up and down, to the right and to the left, and to and fro as a whole, besides which it can rotate about a vertical axis and two horizontal axes. All these motions belong to it as a rigid body, but it is often more than a rigid body and may have internal vibrations which appertain to it as an elastic system. We can count the number of modes of motion, or, as a physicist calls them, degrees of freedom, of a single particle. By adding up the number of the modes of motion of the different particles forming a gas, we can determine the number of modes of motion or degrees of freedom of the gas as a whole. Maxwell remarked that when a gas has settled down to an internal statistical equilibrium, each mode of motion will have on the average a certain energy and this average energy will be the same for all modes of motion. This is a most important theorem in connecting temperature with the other properties of a gas.

It results at once that the ability of a given volume of a gas to absorb energy depends on the number of degrees of freedom per unit volume. The measure of this ability is called the specific heat. It enables us to ascertain how much energy a body in heat equilibrium will contain at a given temperature. If the number of degrees of freedom per unit volume is infinite, the body will be able to take up an infinite amount of energy with a finite increase in temperature; or what is the same, a finite accretion to its energy will not make it any hotter. If we apply a similar argument to a continuous medium, which will naturally have an infinite number of degrees of freedom per unit volume, then this too will have infinite specific heat, and the notion of temperature will not be applicable to it.

Now, Clerk Maxwell was not only the originator of this theory we have just indicated, which is known as the kinetic theory of gases, but also of the theory that light and electricity are transmitted as oscillations of a continuous medium known as the luminiferous ether. This means that the ether can be heated indefinitely without getting any hotter. Since the motions of the luminiferous ether are known as radiation, taking the form of light, X rays, radiant heat, etc., the Maxwell theory of the ether is inconsistent with the existence of any temperature to radiation. Maxwell's theory of light, satisfactory as it is for free radiation in the absence of matter, makes it impossible for light to come to equilibrium with matter in temperature as it is actually known to do in a furnace. Something more and different from the Maxwell theory was needed for the study of the radiation of light, and this something more was suggested by Planck.

Planck observed not only that there is a temperature to radiation, but that the relation between this temperature and the character of the radiation follows a definite law, which is known by Planck's name. In order to justify this law, he supposed that radiation was emitted according to certain small atomic quantities which he called quanta, and this work of his is the first form of the quantum theory of modern physics.

In general, 1900 represents a critical period in scientific thinking. It had not been many years since the most advanced scientists considered that future centuries would be devoted to determining already existing physical theory to further and further decimal places of accuracy. About 1900, however, the quantum theory was beginning to destroy some of the ideas of continuity in the field of radiation. The Gibbsian statistical mechanics was already well on the way to replacing determinism by a qualified indeterminism, and the optical experiment of Michelson and Morley, which showed the impossibility of measuring the velocity of the earth through the ether, had recently become an essential part of the chain of ideas which was to lead to Einstein's relativity.

Einstein's theory of relativity was formulated in 1905, and in the same year he made a critical contribution to quantum theory. He showed that certain of the constants involved in the photoelectric effect, which connects light absorption or emission and electricity, were numerically and dimensionally the same as a famous constant used by Planck in quantum theory. Seven years later, in 1912, Niels Bohr, of Copenhagen, discovered the same constant in the theory of the radiation of the hydrogen atom.

The theory of radiation which was put forward by Bohr was brilliantly although not perfectly successful. It was a curious hybrid in which features of a discontinuous theory were somewhat unnaturally grafted on to a continuous theory like that of planetary orbits. This quantized mechanics had important numerical successes and rather incomplete theoretical unity. By 1925, the year of my talk in Göttingen, the world was clamoring for a theory of quantum effects which would be a unified whole and not a patchwork.

Without being aware of the way in which interest in Göttingen was already concentrating about the difficulties of the quantum theory, my talk in Göttingen, like quantum theory, dealt with a field in which the laws of ordinary magnitudes do not continue down into the range of the very small. As I have said, my talk concerned harmonic analysis—in other words, the breaking up of complicated motions into sums of simple oscillations. Harmonic analysis, for all its many modern ramifications, has a history going back to Pythagoras and his interest in music and the vibrations of the strings of the lyre. There are many ways in which a string can vibrate, but the most elementary and simplest of all is known as the simple harmonic oscillation. The motion of the string of a musical instrument, if indeed it is not simply harmonic, is well known to be the most elementary sort of combination of simple harmonic motions. In fact, for a first very crude approximation, we can treat such a motion as simply harmonic.

Now, let us see what musical notation really is. The position of a note vertically on the staff gives its pitch or frequency, while the horizontal notation of music divides this pitch in accordance with the time. The time notation contains the indication of the rate of the metronome, the subdivision of sound into whole notes, half notes, quarter notes, etc., the various rests, and much else besides. Thus musical notation at first sight seems to deal with a system in which vibrations can be characterized in two independent ways, namely, according to frequency, and according to duration in time.

A finer assumption of the nature of musical notation was that things are not as simple as all this. The number of oscillations per second involved in a note, while it is a statement concerning frequency, is also a statement concerning something distributed in time. In fact, the frequency of a note and its timing interact in a very complicated manner.

Ideally, a simple harmonic motion is something that extends unaltered in time from the remote past to the remote future. In a certain sense it exists *sub specie aeternitatis.* To start and to stop a note involves an alteration of its frequency composition which may be small, but which is very real. A note lasting only over a finite time is to be analyzed as a band of simple harmonic motions, no one of which can be taken as the only simple harmonic motion present. Precision in time means a certain vagueness in pitch, just as precision in pitch involves an indifference to time.

The considerations are not only theoretically important but correspond to a real limitation of what the musician can do. You can't play a jig on the lowest register of the organ. If you take a note oscillating at a rate of sixteen times a second, and continue it only for one twentieth of a second, what you will get is essentially a single push of air without any marked or even noticeable periodic character. It will not sound to the ear like a note but rather like a blow on the eardrum. Actually, the complicated mechanism of the reflection of impulses which is necessary to make an organ pipe speak in a musical manner will not have a fair chance to get started. A fast jig on the lowest register of the organ is, in fact, not so much bad music as no music at all.

It was this paradox of harmonic analysis which formed an important element of my talk at Göttingen in 1925. At that time, I had clearly in mind the possibility that the laws of physics are like musical notation, things that are real and important provided that we do not take them too seriously and push the time scale down beyond a certain level. In other words, I meant to emphasize that, just as in quantum theory, there is in music a difference of behavior between those things belonging to very small intervals of time (or space) and what we accept on the normal scale of every day, and that the infinite divisibility of the universe is a concept which modern physics cannot any longer accept without serious qualification.

To see the relevance of my ideas to the actual development of quantum theory, we must step ahead a few years, to the time when Werner Heisenberg formulated his principle of duality or indeterminism. The classical physics of Newton is one in which a particle may have at the same time a position and a momentum—or, what is not very different, a position and a velocity. Heisenberg eventually observed that under the conditions under which a position can be measured with high precision, a momentum or velocity can be measured only with low precision, and vice versa. This duality is of exactly the same nature as the duality between pitch and time in music, and in fact Heisenberg came to explain it through the same harmonic analysis which I had already presented to the Göttingers at least five years before.

The two main figures of the early days of quantum mechanics in Göttingen were Max Born and Heisenberg. Max Born was much the older of the two; and, although it is unquestionably his line of thought which led to the origination of the new quantum mechanics, the actual initiation of the theory as a separate entity belongs to his younger colleague. Born has always been a calm, gentle, musical soul, whose chief enthusiasm in life has been to play two-piano music with his wife. He has been the most modest of scholars, and it was only in 1954, after grooming other scholars for work which led them to the Nobel Prize, that he himself was awarded it.

Heisenberg, who was at that time in his early twenties, had a less retiring personality than Born and was able to taste success at the very beginning of his life. It must have been a great disappointment for Born to find that his favorite student was drifting off in the direction of German nationalism. The tragedy was made more acute because Born himself was of Jewish origin and Heisenberg ultimately joined the Nazis. This was enough of a tragedy for any one normal life, but we must add to it that later, when Born had retired to Great Britain after the war, his most brilliant student was—Klaus Fuchs.

As I have said before, my Göttingen paper had attracted a certain amount of attention, and hints were dropped to me, emanating from Hilbert, Courant, and Born, that I might be the recipient of an invitation to work at Göttingen some time the next year. In the meantime, Born was coming over to lecture at the Massachusetts Institute of Technology, and I prepared myself to work with him through the intervening months.

When Professor Born came to the United States he was enormously excited about the new basis Heisenberg had just given for the quantum theory of the atom. This theory was an essentially discrete one, and the tools for its study consisted in certain square arrays of numbers known as matrices. The separateness of the lines and columns of these matrices was associated with the separateness of the radiation lines in the spectrum of an atom. However, not all parts of the spectrum of an atom are made of discrete lines, and Born wanted a theory which would generalize these matrices or grids of numbers into something with a continuity comparable to that of the continuous part of the spectrum. The job was a highly technical one, and he counted on me for aid.

There is no point in my going into the technique of a piece of work which not only is highly abstract but was to a certain extent a transitory stage of quantum theory. Suffice it to say that I had the generalization of matrices already at hand in the form of what is known as operators. Born had a good many qualms about the soundness of my method and kept wondering if Hilbert would approve of my mathematics. Hilbert did, in fact, approve of it, and operators have since remained an essential part of quantum theory. They were introduced about the same time by the independent work of Paul Dirac in England. Moreover, they turned out to be useful in tying up another form of quantum mechanics just being invented in Vienna by Erwin Schrödinger with the Heisenberg form of the theory.

From this time on, quantum mechanics went into an active phase of growth in which young men like Heisenberg himself, Dirac, Wolfgang Pauli, and John von Neumann were making new discoveries almost every day. This feverish atmosphere is not one in which I function well, nor did I feel the need of intervening in a subject which was already so competently handled. I did have an idea that the philosophy behind my old paper on Brownian motion could be exploited in quantum mechanics; but the type of difficulty which bothered me and the type of problem where my method was useful were not to become actual for another twenty years. I have returned to this field in the last decade, together with Armand Siegel of Boston University, and I am hopeful that now at last I shall have something useful to say which has not been said by other people.

In all this work, past and to come, I must remember—and the reader must remember—that the task of physics at the present day is not that of carrying out into further and further refinements an existing theory where essentials are well understood. Physics is at present a mass of partial theories which no man has yet been able to render truly and clearly consistent. It has been well said that the modern physicist is a quantum theorist on Monday, Wednesday, and Friday and a student of gravitational relativity theory on Tuesday, Thursday, and Saturday. On Sunday the physicist is neither but is praying to his God that someone, preferably himself, will find the reconciliation between these two views.

5

To Europe
as a Guggenheim Fellow
with My Bride

For the last few years, the meetings between Margaret and me had been a bit too intermittent to suit us. Her teaching and her continued obligations to her own family kept her fully occupied. For my part, my position was not yet sufficiently secure for me to take on the obligations of a married man. Yet the recognition I was receiving from Germany, together with an improved economic status at M.I.T. consequent upon it, now for the first time made it possible for me to look the responsibilities of marriage in the face. Margaret came down to see me at Christmas. I proposed again and was accepted, and we decided to get married and to take the European trip together as our honeymoon.

However, there were difficulties of detail. It was planned that I should arrive in Göttingen in April for the summer semester, at which time Margaret would still be teaching modern languages at Juniata College, in Pennsylvania. She did not wish to resign from the job two months before the end of the school year. For a while we thought of getting married in Europe, but we found that the red tape attendant upon this made it practically impossible. We played with the idea of an embassy marriage or of a marriage at sea by the captain of an American ship. These courses of action also ran into serious difficulties. Finally, we had to admit that the practical and sensible thing was to get married in the United States just before my trip, for Margaret to go back to her teaching work, and for her to join me in Europe during the summer.

Margaret left Boston again for her work at Juniata. In the meantime I found myself very busy and hardly able to think of the new problems of marriage and of the trip. I had an active social life at the time, and on one occasion soon after Christmas the Borns invited a group of us to their apartment to show off a new electric train they had bought for their children in Germany. There was quite a group of scientists and electrical engineers present to witness this occasion—Vannevar Bush, now head of the Carnegie Institution of Washington; Manuel Sandoval Vallarta—the young man who had helped me to translate my article into French—at present vice-minister of education in Mexico, formerly professor of physics at M.I.T., and many others whose names have become household words among those dealing with electricity. When the train was assembled for display and the switch was thrown, the transformer flashed and burned out. It was a considerable time before our combined engineering talents were able to diagnose the trouble. That part of Boston was on direct current, on which no transformer could function.

I was closely associated with Vannevar Bush in my work during this time. Bush was already developing some of the various forms of electrical computing machines which were later to make him famous. From time to time he would call on me for advice, and I tried to do what I could in designing computational apparatus on my own account.

I have already spoken of my work on harmonic analysis, which even at that time seemed to me to be pointed directly toward important practical realizations. Since then these applied realizations have been made; and, as I shall show later, generalized harmonic analysis is an important part of my work even to the present day.

One time when I was visiting the show at the old Copley Theatre, an idea came into my mind which simply distracted all my attention from the performance. It was the notion of an optical computing machine for harmonic analysis. I had already learned not to disregard these stray ideas, no matter when they came to my attention, and I promptly left the theater to work out some of the details of my new plan. The next day I consulted Bush.

The idea was valid, and we made a couple of attempts to put it into working form. In these, my contribution was wholly intellectual, for I am among the clumsiest of men and it is utterly beyond me even to put two wires together so they will make a satisfactory contact. Bush is, among other things, one of the greatest apparatus men that America has ever seen, and he thinks with his hands as well as with his brain. Thus, our attempts in a new sort of harmonic analyzer were quite reasonably successful, and since then they have led to work even more successful.

Ultimately the spring came, and I was about to leave for Germany. I was in a very exulted mood at what I conceived to be the first wholehearted recognition that had come my way, and I am afraid that I talked more of it to the newspapers than was strictly becoming. I felt that I had now got from under the pressure and the indifference of Birkhoff and Veblen. I was more eager to begin the duties of my new position, and I must have been an insufferable young man in my boasting and gloating.

Margaret and I were married in the parish house of a Lutheran Church in Philadelphia. We left at once for a few days' advance honeymoon in Atlantic City, to separate again for the months until Margaret should have discharged her duties at Juniata. She saw me off at the boat in New York. The hotel to which I took her in New York was the old Murray Hill Hotel, which had been for many years the special headquarters of the meetings of the American Mathematical Society, and which was a gloomy old-fashioned marble and porphyry mausoleum inhabited almost exclusively by elderly ladies, around whom flitted the spirits of the not-so-gay nineties.

After this depressing incident I took Margaret to the theater. As luck would have it, we went to see Ibsen's *Ghosts,* the gloomiest of all plays by that most gloomy playwright. These things would not have mattered much if they had only been incidents in a prolonged honeymoon, but as a prelude to a separation of many months they must have been devastating to Margaret's peace of mind.

I arrived in England to see the Devonshire spring already established and the primroses in full bloom. After visiting Hardy at Oxford, where he had now become professor, I went to Göttingen and took up quarters with the landlady of my student days.

I have already mentioned Richard Courant, the young mathematician on whom the administrative mantle of Felix Klein had fallen, as the pope of Göttingen. Courant, who had been amiable on my last Göttingen visit, now turned out to be somewhat hostile. The list of Guggenheim fellows had appeared in the American newspapers, and, as I have said, I had been a bit loquacious concerning my prospective trip. I gave an interview which came to the attention of the all-seeing eye of the Amerika-Institut in Berlin, which went on to dig up the fact of my father's bitter opposition to Germany in the First World War.

While Nazism did not become official in Germany until 1932, there was a strong and bitter nationalist element which had already assumed great power, and which had begun to terrify the more liberal elements in the universities. These universities were of course government institutions, and thus subject to nationalistic pressure. This was also the precise moment at which Courant wanted very much to gain the good will of the United States. The Rockefeller Foundation was deeply interested in European reconstruction. As far as mathematics was concerned, it had picked out the University of Göttingen as the first object of its benefactions. This was entirely natural and right, because Göttingen at that time held an unquestioned first rank as the great center of world mathematics. Later on that year, as I then learned, Birkhoff was to visit Göttingen and to make a detailed report on the project for an improved and enlarged mathematical institute.

I do not envy Courant his difficult position between the upper and the nether millstones. However, it was myself rather than Courant who felt them grind the hardest. Courant's attitude to me became quite cold, and the favors which had been promised me were either denied or granted in such a grudging way that they were not acceptable.

Courant scolded me for my newspaper publicity and was disposed to deny me the assistant and the complete official recognition which he had promised me. Nevertheless, he allowed me to continue in Göttingen in an unofficial way, and after some persuasion found an able young mathematician to aid in the preparation of my lectures, provided I should pay him out of my own pocket.

I was left at Göttingen with a position that belonged neither to flesh, to fowl, nor to good red herring. The humiliation brought me to the edge of a nervous breakdown. Partly as a consequence of this, my lectures were less successful than I could have wished, both as examples of mathematical research and as lectures in the German language. I have no doubt that I would have broken down if it had not been for the loyalty of a few American and English friends, who cheered me up in my blues, went for long walks with me, and attended my classes at a time at which almost all the German students and docents had given them up as a bad job.

Chief among the Americans was J. R. Kline, a Pennsylvania German who was many years later to become the secretary of the American Mathematical Society and the head of the mathematics department at the University of Pennsylvania. He was there with his wife and little boy, and they took me in almost as a member of the family until Margaret came over and lightened their burden for them.

Of my English friends the chief was A. E. Ingham, then a don at the University of Leeds and later a fellow of King's College, Cambridge. Ingham was a shy, almost timid man who had already begun to do distinguished work in the theory of numbers. It is to Ingham that I owe a scientific lead which has carried me to much of my best work.

There were places in my theory of generalized harmonic analysis which I was nearly but not quite able to bring to a definite close. I needed certain theorems, and I found myself proving similar but not identical ones. Ingham pointed out that many similar problems had been solved by Hardy and Littlewood by what they called the method of Tauberian theorems. The study of these is a job belonging to the technique of the mathematician rather than to his repertory of ideas, and I do not intend to try to expound it to the layman. It is enough to say that I made a new attack on this field which was thoroughly successful, and that I not only closed the gap in my earlier work but was able to go on to the simplification of large areas in the theory of whole numbers.

With Ingham and Kline as my friends, I turned my attention to a premature idea of reviving the two old Göttingen clubs, the American and the British Colony which had been the center of my life there in my student days. Kline and I had hoped to improve German-American relations by re-establishing the American colony. Accordingly, we turned for aid to one of the subadministrators of the university.

This subadministrator, who turned out to be a very questionable character, backed us to the limit. He introduced us to a group of young German students, whom I later found to have all the marks of the Nazi about them. Our administrator friend saw to it that our plans got a certain publicity in a local newspaper.

This came to Courant's attention, and he was furious. He squelched the subadministrator with all the fund of contempt which the German professor has for the underlings of the university office. We ourselves got the backlash of his anger, and my very weak position at Göttingen became even weaker.

I had expected that my Göttingen recognition would be a way of getting out from under the continued hostile pressure of Birkhoff in the United States, but now Birkhoff had himself come to Göttingen. He represented the American whose support Courant most wanted.

Courant approached me as an avenue through which he might win Birkhoff's good will. I told him that I had no influence whatever with Birkhoff and that Birkhoff's entire reaction to me was hostile. I kept away from Birkhoff on his visit. I felt that the relations between Birkhoff and Courant were their own business.

Soon after school had closed in the United States, Margaret came over to join me in Göttingen. I fetched her from the boat at Le Havre, and after a few days together in Paris and a short trip to Holland we arrived in Göttingen. It was a pretty sorry and confused state of things into which I introduced her, and it must have been a great shock for a newly wedded wife as yet imperfectly acquainted with her husband. Besides consoling me, she had serious work to do in bringing my landlady to a proper sense of her responsibilities and to a halfway fair treatment of our business relations. The difficult situation in which I found myself as far as my relations with Courant were concerned had gone too far for any repair, but Margaret did her best to help me mend my fences.

Soon after our arrival we threw a belated wedding party for our friends at a well-known Göttingen wine restaurant, where the wine steward did everything in his power to see that the wines we ordered were adequate without being overly expensive. He pointed out that after the first bottle our guests would no longer be interested in the superlative excellence of the wines we ordered and suggested that we order a cheaper wine for the succeeding bottles. Our guests brought us as a wedding present a beautiful tablecloth and set of napkins.

It was not long after that that my parents decided to visit Europe, partly to share in my supposed success and partly to keep a supervising eye over the newly-married couple. This more than doubled my already unsolvable problems. Was I to tell my parents of the rebuff I had received and of the reason for it? As I have said, this lay partly in my father's opinions and in a confusion which the Germans had made of the two of us. It has always been harder for me to be safely wise than to blurt things out, and I told my father what had happened. Naturally, he was much more interested in his personal rebuff than in extricating me from my impossible situation. It was not a very happy week that we spent together in Göttingen, nor was it possible to keep my father and mother from going over my head and attempting to deal directly with the Göttingen people and the German educational authorities.

Margaret and I decided to spend our summer vacation in Switzerland. We went to Bönigen, a suburb of Interlaken, to a little hotel which my sister Bertha and I had already visited on an earlier trip to Europe. Later on the Klines came down from Göttingen to join us at the same hotel. Part of the time we wandered over the foothills of the Alps and some of the time I would play chess with the proprietor, a friendly wine merchant with whom we were on the best of terms. But, suddenly, our stay in Bönigen was terminated by a peremptory summons from my parents, who were passing their holidays in Innsbruck, in Tyrol.

Margaret and I needed this time for those adjustments in marriage which consist primarily in getting acquainted with one another, and which are rendered infinitely more difficult by any attempt at surveillance. On the other hand, through the course of the years, I had become too emotionally dependent on my parents to ignore their summons.

We found Innsbruck delightful, with its walks, its little theater, and its scenery, but my parents were in an irreconcilable mood. Father insisted on my writing an immediate and unconditional protest to the Prussian minister of education. This I knew to be futile, for it was perfectly clear that the minister of Education was the real source of all my difficulties. It was foolish and weak of me to submit, but the habits of years are not easily overcome. It took a long time, even after that, for Margaret to build up in me a certain degree of independence from my parents as an individual and as the head of a family in my own right.

Finally we went to Italy for three weeks of a real honeymoon. We first spent a little time in Bolzano, which had but recently been Italianized from its previous South Tyrolean status as Bozen, and which was not happy under the change.

Then came a brief stay among the dust-covered olive groves of the Lago di Garda.

There followed a visit to the magic city of Venice, with its fabulous watery streets, its treasures of architecture, and the delightful Lido. This Venetian visit would have been pure fairyland if it had not been for the black depression caused by my Göttingen experience.

It was no pleasant experience for Margaret to become involved with the problems of a neurotic husband at his very lowest emotional level. I had become even more of a problem, because my parents had made a policy of glossing over my emotional difficulties, instead of confronting Margaret with the real task she had undertaken in marrying me.

From Venice we went on to Florence and to Rome. Florence in particular seemed to us a city of unbelievable beauty and distinction, which we could appreciate even through the veil of our emotional confusion.

However, the time came when we had to decide what to do with my remaining half-year abroad. For the immediate future, the meeting of the German Association for the Advancement of Science, in Düsseldorf, awaited us. After that, we felt that we had had our fill of Göttingen, and we decided to spend our remaining time in Europe, until January 1927, in the more genial atmosphere of Copenhagen. I had received Harald Bohr's permission to work with him, and I was determined to make up for the blight of my Göttingen visit.

We made a hurried and fatiguing journey to Düsseldorf by way of Switzerland and the Rhine. In Düsseldorf I gave a paper, and I made many new and agreeable contacts with German scientists. In particular, I met a young mathematician named Robert Schmidt, who was an instructor at the University of Kiel. Schmidt had done some important work on Tauberian theorems which, as I saw, was closely related to my own new ideas. We decided to pool our efforts. He pointed out to me in particular that a Tauberian theorem of the comprehensive character, the kind that I had some hope of proving, would be most valuable in number theory, and in particular in the problem of the distribution of prime numbers: such numbers as 2, 3, 5, 7, 11, which have no other factor besides one and themselves.

In the late nineties, two great mathematicians, Hadamard and De La Vallée Poussin, of Louvain, succeeded in proving that the number of primes less than a larger number n was approximately $\frac{n}{\log n}$. Their proofs were thoroughly rigorous and satisfying, but somewhat complicated. Their theorem had been on the point of being proved for many years before they had succeeded in demonstrating it, and the great German mathematician Riemann had come near to establishing it in the third quarter of the nineteenth century. Riemann had made a certain conjecture which he had not been able to establish, but which, if it was true, would lead to a much finer estimate of the distribution of the primes.

To make a long story short, I found my way clear to using my methods to give a much simpler proof of the prime-number theorem and ultimately several much simpler proofs. It was Schmidt who directed my efforts towards this problem, and Schmidt also suggested to me that I might be able to establish or refute the Riemann hypothesis. In this more difficult problem, however, I have always found my efforts completely inadequate.

During my later stay in Copenhagen I made a couple of visits to Schmidt at Kiel. At first he was enthusiastic about my new method, but he gradually began to lose confidence in what I was doing. He threw the work entirely back on my own hands. There were, in fact, some gaps in my proof at that early stage; but they were the sort of gaps which were easily filled up. The repudiation of my work by Schmidt proved a blessing in disguise, for it gave back to me the full control of a piece of research which, if it was not the best that I was ever to do, was certainly close to my best, and which gave me a reputation incomparably greater than that which any of my earlier work had given me.

Courant was at the Düsseldorf meeting, and he tried to get me back again for another term in Göttingen. I told him that a further visit to him would have no point. Margaret and I made a brief trip to nearby Belgium, from which we took an unbelievably fatiguing train voyage to Copenhagen.

To go to Copenhagen by train one had then to take the ferry from Warnemünde to Gjedser. We traveled third class and spent an unhappy night in the red-painted, roughly-beamed, third-class dining saloon on the train ferry. It is a place to make one contemplate all one's past sins and all one's wasted opportunities. The passengers were huddled against one another in an uneasy sleep, and the swinging lanterns cast their oscillating shadows on the floor to the tune of the rocking of the boat and the creaking of the timbers.

When we got to Copenhagen we were nearly dead, and we slept for a whole day. Then I looked up Harald Bohr and pre-pared for some months of research. We saw a good deal of the brothers Bohr. I remember that at the apartment of one of them, I believe it was Niels, there was a plaque of the two as children which gave them an undeniable peasantlike appear-ance, which they had lost in the course of the years. One of the other guests, a lady who was professor of classics at the University of Copenhagen, and who smoked big black cigars incessantly, told us that some friend had commiserated with their mother for having two such dull boys for children. In view of the fact that Niels Bohr has become the national hero of Denmark because of his scientific work and lives in a palace donated by one of the great Copenhagen breweries and that

Harald was certainly the greatest mathematician ever to live in Denmark, this story now seems more than a little ridiculous.

Copenhagen was a delightful city which combined the intellectual amenities of a world capital with the hominess of a small town. In the intellectual world everyone knew everyone else, and there was an atmosphere of friendliness pervading the whole of life.

The Bohrs were charming to us, as was their colleague, Professor Nørlund, whom I had already met in Strasbourg. Nørlund was a tall, handsome, bearded man, who had gone from pure mathematical work to the headship of the Geodetic Survey of Greenland, and whose house was frequented by bluff Arctic sea captains. Mrs. Nørlund retained the beauty which had so impressed me at Strasbourg. She was most cordial to us. We had already decided to learn Danish, and were taking Danish lessons with a high-school teacher who had spent some time in the United States. Mrs. Nørlund supplemented our Danish instruction by reading with us Andersen's *Fairy Tales* in the original. The beauty of these fairy tales in the sweet, intimate Danish language was brought out to the full by the charming way in which she read them.

Besides my work on Tauberian theorems and number theory, I made one or two new starts in Copenhagen on important points in harmonic analysis. Copenhagen was a rest and a refuge after the turmoil of Göttingen.

I have said that we left Copenhagen for a brief trip to Germany, where my wife visited her relatives and I worked with Robert Schmidt. After that we returned to the extended festivities of the Danish Christmas and New Year season. Two weeks are devoted to nothing but parties, and the crowds take delight in milling up and down the narrow business street of Strøget.

The time had now come to go back to the States. We returned by way of England. Margaret and I took over the room of some American friends of hers who had been studying in London. I found the mild winter a fine occasion to talk over my work with Hardy and take advantage of his criticisms. Then we returned to the States after a calm winter voyage and spent a day or two with my sister Constance before we started house hunting.

6

1927—1931.

Years of

Growth

and

Progress

We stayed a few days with my sister on Pleasant Street, in Belmont, where there had been a blizzard just before we arrived. I remember how soothing to me was the muted sound of the slapping of heavy chains on the snow. Then we started house hunting. We found an apartment just across the Arlington line. I began to try to adapt myself to a life of handyman domesticity, for which I had no particular qualifications. I made good after a fashion as a furnace tender and furniture varnisher, but this was never my métier.

Dirk Jan Struik, a Dutch mathematician whom I had met at Göttingen and who was now a new appointee at M.I.T., had already come over some months before and was immersed in his work. He fitted into the familylike environment of our department very well, and I began to study his field of differential geometry. We started work together on an attempt to apply his ideas to differential equations and, in particular, to the Schrödinger equation of quantum theory. We were off the main line of progress, but we did obtain some interesting theorems. Our work was not the sort which makes a big splash in the literature, but it was the sort which is rediscovered years later, and which has a permanent but limited interest.

The Struiks and ourselves spent the summer in our beloved town of Sandwich, in New Hampshire. We used as our headquarters a boarding house which had long been known to us. Margaret was pregnant and terribly uncomfortable with poison ivy rash, so she could not accompany Struik and me on our hikes. The two of us, however, ranged the nearby mountains and made a larger expedition together to the Presidentials. From this we returned bearded like the pard. Struik promptly shaved off his beard, which had made him look as if he had been painted by Rembrandt, and Margaret started to trim mine down by degrees, until it attained the exiguous proportions of the beard which I continue to wear to the present day.

We had many friends among our country neighbors. These included Clare George, a well-to-do, eccentric, mannish spinster, who affected trousers at a period when these were not yet in vogue for women, and who used to puff a furtive cigarette when she was alone with my wife. We saw her often at the home of our friends, the Corlisses. Louis Corliss is a Cornell engineer who had worked for the Sperry Gyroscope people in the early days, until a series of deaths in the family and ill health due to overwork had led him to choose the life of a farmer on his family farm rather than the confusion of modern engineering and industrial life. He was a widower, living with his grandmother and his mother. The whole family was charming to us and has continued to be among our close friends. Grandma Corliss died some twenty years ago, and Corliss married the nurse who attended to her in her last days.

Their daughter, Janet Corliss, has become my trusted secretary during the summer months and has assisted me with the preparation of this manuscript.

Clare George and Louis Corliss's mother were well aware that I loved the Sandwich region and that Margaret was coming to love it. We felt that we wanted our prospective children to have the advantages of country life that had been granted in one way or another to each of us. Our friends made themselves busy scouring the neighborhood for a suitable summer home. They found one on a knoll on the Bear Camp Pond road. The house was uninhabited. It had been only recently inhabited, however, and was in good condition. When Margaret and I made the circuit of its weed-grown lawns and peered through its cobweb-covered windows into its graciously proportioned rooms, we knew we had found what we wanted.

We sought out the names of the caretaker and of the lawyer who was entrusted with the property. The region had been going downhill from the Civil War to that time, and real estate prices were at a dead bottom. The price named, if not within our immediate reach, was not outside what we could hope to pay within a few years. We agreed that it was the summer place for us.

Soon after that, my parents came up to visit us at our boarding house. We showed them the country place we were thinking of buying, and they were enthusiastic about it and helped us to purchase it. Ever since, our vacations in the White Mountains have meant for us the relaxation we need from the strenuous life of M.I.T., and also a chance to give our children the experience of the country and the freedom of living which we feel to be the birthright of every child.

As a matter of fact, Margaret and I needed the place as much as we hoped our prospective children would need it. Teaching at a university is a very strenuous job, but teaching combined with research is a full load for any man. Much of my research depended on free exchange of ideas with other men; but there always comes a time when this preliminary work is done and when I have had to spend my main attention on writing up the work in compact and acceptable form. This writing can be done best when there are no distractions and when my life is a simple alternation between concentrated intellectual effort and the completely non-intellectual pleasures of roaming the countryside, meeting my non-professional friends, and swimming and basking on the beach.

There are many people who believe that the summer vacation of a college professor is a very special sort of junket, a pleasure granted to the intellectual in return for his smaller salary and his unexalted position in the American scale of social values. Nothing could be further from the case. Severe work of a research nature drains one dry, and without an ample opportunity to rest as intensely as one has worked the quality of ones' research must go down and down.

I do not mean to make the claim that only intellectuals need long vacations. I am quite certain that the continuous pressure of industrial work and the fragmentary vacations given as a surcease for this work are responsible for the early aging of many of our best minds in industry. This condition has become particularly acute since the war, for we have acted on the hypothesis that in times of stress it is a sort of treason to relax. I am convinced that our policy of continued tension is foolish and that it fails to serve the end of the best use of our human resources.

My older daughter, Barbara, was born during the next academic year, 1927–28, and I began as a very clumsy pupil in the art of baby sitting and of hanging out a long signal hoist of diapers.

The next summer, we settled on our new estate with our new baby. My father had given me his superannuated Model T beach wagon, and he had made several trips with me to bring up the necessary furniture. In those days we were without a telephone, without electricity, and even without a stove. We prepared Barbara's formula in the fireplace and did our rudimentary cooking there until such a time as we could get a second-hand two-burner oil stove. To the present day we remain without running water, although we have found a very satisfactory substitute for this in the form of a force pump and gravity tank.

We used to take the baby down to the beach of Bear Camp Pond, where she, like our second daughter, Peggy, practically grew up in the water. The beach was semipublic and was frequented by a large number of neighbors with children of all ages. I initiated these children into the habit of taking long mountain hikes, and now when I go to the beach, I see there the children of these children.

We had the Klines as our guests that summer. They were charmed by the region, and ultimately decided first to rent and then to buy a summer cottage there. I have said that the region was about at its lowest economic point at the time that we bought our house. From the agricultural standpoint, it has perhaps lost further ground, but it has become a popular summer home for a very congenial middle-class group, among them a number of university people. Some of these, indeed, have since retired, and spend a large part of the year in those restful surroundings. At present it would be easy to recruit the faculty of a very fair university from our neighbors without bringing anyone in especially for the purpose.

The improvement in my scientific status went on apace. My new personality as a married man made it possible to allay some part of the hostility with which I had been received in mathematical circles. Nevertheless, the many barriers against me were still up. Birkhoff had made his prejudice a matter of principle, and saw to it that many academic offers which otherwise might have been open to me were diverted elsewhere. As this was the period during which the status of the college professor was improved greatly over the country as a whole, and in which most of my colleagues were preparing to save for their old age, this deprivation was serious for me. Tech indeed kept on advancing me steadily, notwithstanding the absence of outside offers, but it is an unquestioned fact that the existence of such outside offers would have improved my position in one way or another.

In default of American offers for an improved position coming through the normal channels, I began to look around and to see if I could not do something for myself elsewhere. The British universities and the universities of the British colonies operate under the legal provision that if any vacancy occurs it must be advertised and the applications of all candidates must be considered at least in a formal way. This requirement is not taken too seriously, and in many cases a decision has already been made for all practical purposes at the time the vacancy has been advertised. These advertisements appear on the back pages of *Nature* and other British intellectual publications. I sent in my name for one vacancy at Kings College in London and for one in Australia, but of course nothing happened. Nevertheless, the practical evidence that I was trying to better myself was of considerable benefit to me as far as my M.I.T. advancement went.

It was about this time that there was a summer meeting of the American Mathematical Society at Amherst. Margaret went with me, and we both thoroughly enjoyed the occasion. The important matter for me was that I saw there a great deal of my friend, J. D. Tamarkin, another Göttingen acquaintance, and that he seemed favorably impressed with my research. He became my enthusiastic and sincere backer, and it was more through him than through anyone else that my American recognition began to take serious proportions.

Tamarkin was a brilliant mathematician whose origins were

in those days before the First World War, when life in Russia
was very good for those above middle-class standing. In America he attempted to carry out the open-handed lavish hospitality which belonged to prewar Russia. He had escaped from
Russia at the risk of his life and had won an enthusiastic reception from Professor R. G. D. Richardson, of Brown University.
Tamarkin's mathematical standards were of the highest, and he
welcomed my work at a time when those of a purely American
tradition did not think very much of it.

During these years, Hardy made a series of visits to the
United States. He also thought well of my work, and between
him and Tamarkin I began to be heard of in this country, but
I was never able to forget that the people to whom I owed
the greatest part of my recognition were not Americans.

The years after I came back from my trip of 1926 and before
my trip of 1931–32, were of course the years of Coolidge prosperity and of the depression. Even in our relatively protected
academic life we felt the strong impact of both phases of
national and world existence. As I have said, Harvard salaries
had been raised during the boom years to a very considerable
extent; and although Tech salaries lagged, there was still the
expectation on the part of many that they might gradually
climb, if not to Harvard levels, to a reasonable imitation
thereof. As a consequence, many of my colleagues at both institutions were talking stock market and behaving like capi-
talists. You could not get a group of five college professors
together without hearing a comparative evaluation of the popular stocks of the day. One or two of my younger colleagues
devoted more attention to the course of their investments than
to their academic work.

I never fully believed in the boom, though I was quite well
aware of its consequences on our lives. Too many of the values
were paper values which, as I even then saw, could vanish over
night. The farmers went in for those silver-fox farms which

even the slightest slump would deprive of a market. Some of my colleagues tried to supplement their incomes by breeding fancy kinds of dogs and Siamese cats, and these were subject to a similar disadvantage. In the same category of will-o'-the-wisp prosperity were the land boom in Florida and the vogue of Steuben glass and antique furniture. We never had enough money to go in for any of these things, and, frankly, I never felt the temptation. Thus, I was prepared for the boom to end in a crash in which the loss of a set of paper values was bound to bring as a consequence the loss of an entire structure, which had flourished like the green bay tree.

The paper values of a monetary nature involved a whole series of paper moral values. I was astonished and shocked at the way in which a great national magazine gave its columns over to a panegyric of the Swedish match king, Ivar Kreuger. What I was less prepared for was the acceptance of the same moral values in academic circles. I hoped and prayed that the slump would come early before it could build up into the complete failure that generally follows the burst of a promotional bubble. I talked this over with my good friend Phillips and was surprised to find that, for all his native skepticism and personal shrewdness, he was hopeful that the boom was here to stay. Behind this feeling on his part was a long experience as a youth in the ruined South after the Civil War and a fear that we might be heading for another such period. To a large extent, his optimism was a whistling to allay his own fears, but at any rate he did not share with me the hope that a mild depression might turn us away from the fleshpots of Egypt and to a greater evaluation of moral and intellectual matters. When the slump came, I saw that he had been justified in his fear that a collapse would not only destroy commercial values, but moral values as well.

The slump was bound to affect everyone, but we academic people had the best of it, for a while. To some extent, prices went down, and while our expectation of rapid advances in salaries blew up in our faces, many of us, like myself, had tenure; and those of us who did not have it had a moral expectation of it which was not easily canceled. At any rate, we did not suffer from the epidemic of suicides which took place among businessmen, and a high window was no particular moral hazard for us.

A college professor who is really interested in his work is considerably insulated from the vicissitudes of the world about him. At the present day, when science is an object of general attack, and when many of us have serious doubts whether our civilization really is viable, the protection of our isolated position has largely vanished.

In those days—the late twenties and early thirties—however, while we may have doubted many things, we did not have fundamental doubts of the long-time recovery possibilities of the world in which we lived. Thus the Sacco-Vanzetti case, the phony boom, and the almost equally phony depression that followed it, drove us more and more into ourselves and into our real function, academic work.

My work, as I found it at this time, was research and the initiation of research students into their proper activity. As my work and reputation developed, I began to get graduate students. There was an M.I.T. professor whose son was doing research in the Harvard mathematics department for the doctor's degree and who wanted to work with me. We found it possible to make an arrangement between the two departments by which his degree should be at Harvard while I should be the director of his thesis.

My first M.I.T. doctoral candidate was a young man by the name of Carl Muckenhoupt. I gave him a thesis topic in Harald Bohr's theory of almost-periodic functions. These functions had been studied by Bohr as purely abstract mathematical entities, but I saw how they could be used as effective tools in the qualitative and even the computational study of vibration problems. The Muckenhoupt thesis represented another link in the synthesis between pure and applied mathematics which I conceived it was my function to make.

Important as the Muckenhoupt work was in my development, there were two pieces of work that came along slightly later which turned out to be even more important. These represent doctoral theses which, by a peculiar coincidence, were done under my supervision by two students from the Far East. Their names were Yuk Wing Lee, from China, and Shikao Ikehara, from Japan.

The way in which I came across Lee is interesting. My Dutch friend, Struik, had found some summer work with the Bell Telephone Laboratories in connection with the analysis of electrical circuits. This immediately led me to consider whether I might not have a different approach to the same field by the use of Fourier series. My idea continued to look good to me under further examination, and I asked Vannevar Bush whether he could lend me a good student in electrical engineering to do a thesis under me. He was only too glad to do so, and suggested Lee, who was then living at the parish house of one of the Boston churches. Lee readily accepted the offer and we went to work together.

Lee and I have been scientific colleagues now for about a quarter of a century. From the beginning, his steadiness and judgment have furnished exactly the balance wheel I have needed. My first idea of an adjustable corrective network would have worked, but at the cost of a great wastefulness of parts. It was Lee who saw how the same part could be used to perform several simultaneous functions and who in that way reduced a great, sprawling piece of apparatus into a well-designed, economical network.

It was also Lee who found a possible purchaser for our invention, in the form of a research corporation allied to the

moving picture industry. It was Lee, above all, who went down
to Long Island City and spent many months of patient work in
developing our apparatus, computing the size of the parts,
constructing a good working model which functioned the first
time up to the degree to which we had predicted its functioning, and selling the idea to our clients.

Unquestionably those were thin times for the electrical end
of the moving-picture industry, which recently had been created to take care of the problems of the talking film. Our rights
reverted to ourselves, and we had to look for a permanent purchaser. It was Lee who found such a purchaser in the Bell
Telephone Laboratories and who saw the work through the
tedious stages of patenting.

Here I must remark that the public that is interested in
inventions but has had no direct experience of the Patent Office
can have no adequate idea of the utter boredom of seeing an
invention through the necessary stages of search and documentation. In the first place, it means nothing merely to patent an
invention. Any teeth that the patent may have—and they are
few enough when the patent is held by a private individual
with no means—depend on a detailed legalism of the phrasing
of claims and specifications, which have very little to do with
the actual merits of the invention. Here the patent lawyer can
be of great help, but he can be of very limited help unless he is
backed by the peculiar understanding of the invention which
only the inventor himself may be expected to have.

The result is that the inventor proceeds without any transi-
tional stages from a game of ideas to a game of words. The
more he loves his invention for its own sake and the more that
he wishes to develop it, the more he finds himself frustrated
by the unreal world of the Patent Office, in which he is forced
to live for a term of months or years.

At the end of these months or years, Lee and I found ourselves possessed of a salable invention—indeed, of an invention

which we had actually sold before the final stages of its patenting were completed. But then we met the further frustration that all this effort we had made went into a paper patent—that is, into a patent which the Bell people never intended to use but simply to hold *in terrorem* against against competitors.

The Bell people never did use our invention, from the time it was on the books of the Patent Office to the time its patent expired. Nevertheless, as our document was near the end of the seventeen years, which are to a patent what seventy years are to a man, we found that certain radio and television firms began to show a great curiosity concerning the new invention, as if in fact they were about to incorporate it into the sets they made. Since our rights have expired, we have never shown the curiosity to find out whether these engineers have in fact followed our ideas to the point of execution or our baby was stillborn. This does not prevent us from having a shrewd idea that, whether or not the invention was ever made for purposes of sale in the form in which we wrote it up, it is still exercising an influence on the philosophy of the art.

I made every effort to place Dr. Lee in the American electrical-engineering industry, where he would have been a valuable man. At that time, the engineer from Eastern countries did already exist in the United States, but he was a much rarer bird than he is today. The sales resistance I had to meet was more than we could overcome, and Lee went back to China to seek a job, first in industry and then in academic life. I shall have much more to say about him in later chapters.

At the same time at which one Oriental student, Lee, was developing the consequences of some of my engineering ideas, another, Ikehara, was perfecting my methods in prime-number theory. Landau, the chief German exponent of prime-number theory, was at first hesitant to accept our results, but ultimately he and his colleague, Heilbronn, wrote papers taking this work still further. The result was to remove a difficult branch of mathematics from the latter years of graduate work and to make it available even in an advanced undergraduate course. In recent years, the Scandinavian school of mathemations has gone even beyond where we left our work and has made prime-number theory elementary in a certain very technical sense.

I have already mentioned Vannevar Bush as a great inventor of electromechanical devices, and his chief work was along the line of the development of high-speed computing machines for solving problems in the field known as technique of differential equations. Differential equations are concerned with the relations between various measurable physical quantities and their rates of change in time and space. These physical quantities can be currents or voltages or the angles of rotation of shafts or quantities of still different sorts. Bush's device now tried one set of quantities as their bases and now another, but the form to which his apparatus gravitated was a sort of meccano set, the differential analyzer, in which various quantities could be represented, let us say, by the rotation of a shaft, and in which these quantities could be added, multiplied, divided, and operated on in still other ways. Above all, where the total sum of a quantity was wanted, these quantities could be read off by the device known as an integrating disk.

None of the individual parts of the Bush machine were completely new in conception, but the technique by which they could be combined, and in particular by which the power to move the apparatus could be put in locally in such a way that the machine would not stick, represented an improvement in technique going far beyond anything conceived before. Bush's machine was successful where Babbage's earlier machine had failed, precisely because of a brilliant use of engineering facilities and engineering ideas not accessible at the time of Babbage.

In Bush's machine numbers were represented as measured quantities rather than as sequences of digits. This is what we mean by calling the Bush machine an analogy machine and not a digital machine. The former measures. The latter counts. The physical quantities involved in the problems which the machine was to solve were replaced in the machine by other physical quantities of a different nature but with the same quantitative interrelations.

At that time the digital machine, which is an improved and automatized abacus, was confined to various forms of the desk computing machine.

It was an essential part of Bush's machine that the variable in which all changes were to take place was time. This became exceedingly significant when Bush asked me for advice on how to make his machine take care of partial-differential equations, in which time rates of change and the space rates of change are united by equations.

When Bush asked me this question, I realized that the main problem of the partial differential equation was that of the representation of quantities varying in two or more space dimensions, such as, for example, the density of a photographic negative, which varies up and down and right and left.

Once the problem of representation of functions of several variables was clearly stated, it became desirable to represent these too as something changing in time rather than in space. Here it appeared to me that the new and developing art of television gave the necessary clue. In television, a picture is conveyed not by pieces of silver of various opacities placed simultaneously on a film but rather by a dot of light running over the various rows of a grid point by point, and the whole grid row by row. This process, called scanning, is now familiar to anyone who has the least curiosity as to how his home television set works.

In fact, I was convinced that the scanning technique would prove socially more important in computing machines and their close relatives than in the television industry itself. The future development of computing machines and control machines has, I believe, borne me out in this opinion.

In representing a quantity as a single locus in a television screen, we may follow two different techniques, one derived from the analogy machine and one from the digital machine. Each spot on the television screen may determine either a quantity of light which is measured by its intensity or a sequence of digits such as we use in writing down a number. The combination of these quantities to represent the situation present in a partial-differential equation may accordingly be either a combination of intensities or a combination of the digits of several numbers. It seemed to me even at that time that the latter method of representation would be more suitable for partial-differential-equation machines, because of the fact that, with electronic apparatus, digits may be combined more accurately and more expeditiously than amounts of light. I need only say that the actual development of the computing-machine technique has proved the correctness of my surmise and that the high-speed computing machines of the present day follow very closely along the lines which I then suggested to Bush.

The emphasis which I then put and still put on speed in mechanical or electrical computers has fully justified itself. We can put more distinguishable numbers on a square grid than we can on any one line of the grid, and the number of operations through which we must go to represent the process of solving a partial-differential equation is simply enormous. Without stepping up the speed to a tremendous extent, the partial-differential equation machine would have been so slow that it would have become useless. In general, the computing machine is a competitor of the human computer; and when all is said and done, its advantage over the human computer lies primarily in its speed. This subject, to which I shall revert later in the present book, represents the first step toward the origin of the extremely high-speed computing machine of today as well as toward the closely related machines of the automatic factory.

At about the same time, Bush was engaged in the writing of a book on electrical-circuit theory. Here some of the work which I had done on generalized harmonic analysis became of considerable practical value. He called me into consultation for advice concerning many of the chapters and also asked for a supplementary chapter on Fourier methods. We enjoyed the collaboration greatly, and both of us have spoken often of the fun we had in working together. Bush soon left his theoretical work for an administrative career. This was a step in the formation of a new academic setup which had long been overdue.

When I arrived at M.I.T., in 1919, Richard McLaurin was president. His was a name to conjure with, and he had added enormously to the position of the Institute at home as well as overseas. However, he died within a term of my arrival and left incomplete much of what we had hoped he would achieve. In particular, the science departments, including mathematics, and the cultural departments, such as those of English and history, were still conceived as service departments for the main center of life in the Institute, which was engineering.

After McLaurin's death we went into the doldrums for eleven years. For part of that time we were ruled by committees of the faculty, which could do very little because of their avowed temporary character, and for another part by President Ernest Nichols, who, however, was already in ill health when he came, and who retired and died before he could leave much of his impress on the school.

Finally, President Wesley Stratton was appointed, on the strength of the good works he had done as the head of the Bureau of Standards. Like Nichols, however, he came at a period when his best work was over, and he did very little but prolong the interregnum. Until 1930, there was no ruler of the Institute with a sure touch, a clear policy, and an unquestioned vigor and understanding.

Then Karl Taylor Compton was appointed president. He had been a distinguished professor of physics at Princeton, and he combined complete integrity, a long view for the future of the Institute, and a still untapped health and strength. He inspected the physics department on behalf of the Corporation and saw clearly what none of his predecessors had had the opportunity to realize, that a strong engineering school must be at the same time a great school of science.

The role of mathematics at the Institute had changed greatly since the days after the First World War, when I came to M.I.T. At that time, mathematics was something which was chiefly needed to educate our students up to the point where they could handle the engineering which was their main object in life. Physics and chemistry, too, had not yet emerged completely from their status as service departments, whose main purpose was ancillary to the chief task of the Institute, which was to train engineers. When a branch of physics or chemistry had reached a sufficient importance in its own right for a specific course in it to be made a new branch of engineering, this course was set up as an independent engineering course. This had been the history of our electrical-engineering department and of our chemical-engineering department. Now for the first time the Institute had begun to be aware that direct research in mathematics and the sciences was of engineering importance in its own right and that we should devote ourselves explicitly to the training of scientists in these fields as well as to the training of engineers.

In the mathematics department in particular, this removed a great blight under which we had been suffering. Our research began to be recognized as an essential part of our function at the Institute rather than merely as a way to keep us on our toes so that our routine teaching could be fresher and more authoritative. The Institute began to follow the example of the great universities in recognizing us as mathematicians rather than solely as mathematical-routine teachers. This does not mean that we gave up or that we could give up the service work so necessary at an engineering school, but it did mean that we had begun to come into our own, on a status comparable with that of the members of the engineering departments.

President Compton was accessible, modest, sincere, and lovable. With his appointment, the Institute was once more in strong hands, and the line of progress, which had been interrupted by McLaurin's death, was resumed.

Bush's advancement was part of the same movement that brought Compton to the Institute. He was a splendid administrator, taking as his particular field the laboratories of the Institute rather than the personnel. He relieved Compton from much of the detail inseparable from the running of a great school.

One part of Compton's policies was to bring faculty salaries to a level comparable with those already reached at Harvard, Princeton, and the other major universities. Later on, the war and the vicissitudes of the post-war period prevented us from catching up as completely as we might have wished with other schools of the same level. However, the intention was there, and much was done to realize it.

I was a personal beneficiary of the new regime, both as far as salary was concerned and as far as the opportunity to see my hopes for a research mathematics department come into being. I received my promotion to the rank of associate professor, and from that time on my status was assured.

It was thus at the time of M.I.T.'s new impetus of life and Bush's greatest development as an electrical engineer that he had offered me Lee as a graduate student. This was one of the finest things Bush has ever done for me, and I am eternally grateful that he turned Lee in my direction.

About this time we began to find ourselves pleasantly occupied with the visits of a number of scientific colleagues from Europe. In connection with circuit analysis and the electrical-engineering aspects of my work, I saw a good deal of Richard Cauer and his wife, who had come over from Berlin. However, the scientist with whom I had the most interesting and profitable contacts was Eberhard Hopf. He had come over from Germany to Harvard, largely to study with G. D. Birkhoff.

Hopf's interests had been in celestial mechanics, and the new work of Birkhoff. The ergodic theorem, which finally gave the proper form to the ideas of Willard Gibbs, was exactly along the line of Hopf's interests. This piece of work, by the way, was a remarkable tour de force, as Birkhoff had gone into the subject cold, with no particular previous knowledge or interest in the Lebesgue integral. However, he had managed, by his own powers, to extract one of the leading theorems which has dominated the theory of Lebesgue integration ever since.

I was much interested in Lebesgue integration and probability theory, so Hopf and I had a great deal to talk about. However, the best of the work which he and I undertook together concerned a differential equation occurring in the study of the radiation equilibrium of the stars. Inside a star there is a region where electrons and atomic nuclei coexist with light quanta, the material of which radiation is made. Outside the star we have radiation alone, or at least radiation accompanied by a much more diluted form of matter. The various types of particles which form light and matter exist in a sort of balance with one another, which changes abruptly when we pass beyond the surface of the star. It is easy to set up the equations for this equilibrium, but it is not easy to find a general method for the solution of these equations.

The equations for radiation equilibrium in the stars belong to a type now known by Eberhard Hopf's name and mine. They are closely related to other equations which arise when two different physical regimes are joined across a sharp edge or a boundary, as for example in the atomic bomb, which is essentially the model of a star in which the surface of the bomb marks the change between an inner regime and an outer regime; and, accordingly, various important problems concerning the bomb receive their natural expression in Hopf-Wiener equations. The question of the bursting size of the bomb turns out to be one of these.

From my point of view, the most striking use of Hopf-Wiener equations is to be found where the boundary between the two regimes is in time and not in space. One regime represents the state of the world up to a given time and the other regime the state after that time. This is the precisely appropriate tool for certain aspects of the theory of prediction, in which a knowledge of the past is used to determine the future. There are however many more general problems of instrumentation which can be solved by the same technique operating in time. Among these is the wave-filter problem, which consists in taking a message which has been corrupted by a simultaneous noise and reconstructing the pure message to the best of our ability.

Both prediction problems and filtering problems were of importance in the last war and remain of importance in the new technology which has followed it. Prediction problems came up in the control of anti-aircraft fire, for an anti-aircraft gunner must shoot ahead of his plane as does a duck shooter. Filter problems were of repeated use in radar design, and both filter and prediction problems are important in the modern statistical techniques of meteorology.

In the fall of 1929, I received an invitation to lecture on my own research at Brown University. Dean Richardson invited me, but the spirit behind the invitation was Tamarkin whom I have already mentioned. Richardson was a dry, friendly Scotsman from the Maritime Provinces who gave Tamarkin a home in the United States, much to the advantage of Brown University. I traveled once a week to Brown, where I had found a most cordial reception. Tamarkin, together with Mrs. Tamarkin, who had by now come to join her husband, was my principal host.

The Tamarkins had carried their expansive style of living into an America where the habits of the country and the difficulties of the servant problem made this sort of a life almost impossible. Mrs. Tamarkin struggled courageously in the restrictive environment of Providence to feed her husband's habitual need for good food and drink, but in the course of this effort she wrecked her health by overwork while her husband continued to overload his damaged heart. Mrs. Tamarkin died ultimately of an attack of phlebitis, but her husband attributed it to overwork and reproached himself. Before her death he had been the soul of jollity and good cheer, and he continued to offer unstintingly to the younger mathematicians from the store of his great knowledge and sympathy. Yet he never was quite the same man again, and a few years later he too succumbed to the strains which he had imposed on his heart.

Peggy, my second and last child, was born in that year, and I went directly from my vigil at the hospital to my Brown lecture. With two babies in the family, I became much more of a family man and Margaret much more occupied with family duties.

I was frightfully busy at the time working up my definitive paper on generalized harmonic analysis. This appeared in *Acta Mathematica*, a Swedish journal of great international prestige.

The paper was practically a small book. It was Tamarkin who urged me to write up this work in definitive form, and it was he who criticized every stage of my manuscript and proof, to its great advantage. I think my papers satisfied Tamarkin to some degree, and it was certainly as a result of his backing that I soon received an invitation to write for the American journal, *Annals of Mathematics*, a paper of similar comprehensiveness concerning Tauberian theorems.

These papers assumed the proportions of small books. As to the *Annals of Mathematics* paper, it was actually published as a separate memoir. In my later writing I have often wished that I had the continued advantage of Tamarkin's selfless criticism.

My research at this time received a ready reception in Russia and was in close relation with the work of some of the Russian mathematicians. I had long had a peculiar sort of contact with the leading Russian mathematicians, although I had never met any of them nor, I believe, ever been in correspondence with them. Khintchine and Kolmogoroff, the two chief Russian exponents of the theory of probability, have long been involved in the same field in which I was working. For more than twenty years, we have been on one another's heels; either they had proved a theorem which I was about to prove, or I had been ahead of them by the narrowest of margins. This contact between our work came not from any definite program on my part nor, I believe, from any on theirs but was due to the fact that we had come into our greatest activity at about the same time, with about the same intellectual equipment.

Four and a half years without travel abroad had put me in a mood for renewed foreign contacts. Since the International Mathematical Congress at Strasbourg, two more Congresses had passed in which I had not participated. The Congress of 1924 took place at Toronto, but, as I have said, I had devoted that summer to my trip abroad with my sister Bertha. That of 1928 occurred too soon after my Göttingen trip to make my attendance possible, particularly as it was the year of the birth of my elder daughter, Barbara.

By 1932 I found the urge to attend another Congress too strong to resist. I planned to spend the academic year of 1931–32 studying at Cambridge and to participate in the Zurich Congress the following summer. I received generous assistance, in the matters of both leave and finances, from the Massachusetts Institute of Technology, so that Margaret and I found it possible to venture a European trip together with our young family of two.

7

An
Unofficial
Cambridge
Don

We spent the summer as usual, resting and hiking. I took rather
longer walks with the children of the valley than I had before,
since they were now old enough to stand the severe trip over
the Presidential Range. Shortly before we left to take the boat
at Montreal, I went with a few of my young friends for a quick
trip to Mount Chocorua, where one of them had the misfortune
to turn his ankle. The problem of getting him down was considerable, since he had to stand behind me and rest his weight
on my shoulders for two grueling hours. The result was that I
was tired and had a bad cold coming on when the time came
for us to make the trip.

This was not the worst of it. Peggy, who was about one and
a half years old at this time, had managed to develop a slight
fever; and although the local doctor thought that a sea trip

would quickly put her in shape, we were none too confident. Thus, we were rather under the weather when we took the train at Meredith for Montreal and our embarkation.

After a difficult night at a hotel, we just managed to embark in the morning. Then our illnesses really gripped us. Fortunately, the Canadian Pacific boat on which we traveled had an excellent doctor, who had himself taken to the sea to recover from a complicated and crippling illness, as well as a first-rate Scottish trained nurse and a couple of ex-N.C.O.'s of the Royal Army Medical Corps. Peggy and I went immediately to bed, and while the nurse took care of her in our cabin, I went through a bout of septic sore throat under the care of one of the medical stewards.

When I had recovered enough to drag myself around the deck, I found that Peggy was still no better. The doctor came and listened to her lungs with the stethoscope. As soon as I saw that he had settled down to the detailed examination of a particular spot on her chest I knew we were in for trouble. It was bronchopneumonia all right, and for about half of the voyage we were not at all certain of Peggy's recovery.

The doctor consulted with another doctor who was aboard as a passenger, and between them they hit on a course of treatment. By the time the voyage was half over the worst was passed. We wirelessed ahead for an ambulance to meet the boat at Tilbury docks to take Peggy to the local hospital, and settled down to enjoy the remainder of the trip as much as we could with a considerable burden of worry still on our minds.

In due course we landed, and Peggy was taken to the hospital, where it was considered advisable for her morale and for that of the hospital that we should see as little as possible of her until we had found quarters at Cambridge and were prepared to take her back with us.

We stayed for a few days at an excellent hotel run by the

Port of London Authority, where I had the pleasure of meeting the personal representative of that body—the captain of the port and harbourmaster of London. He was much interested in engineering, and especially meteorology. He told me what a lonely profession his was, with a small group of perhaps less than ten men in the whole world with a comparable job; and how badly he had felt during World War I to have been shut off from his opposite numbers at Hamburg and at Antwerp. He said that, in fact, under normal conditions, London, Antwerp, Rotterdam, Hamburg, and Bremen all constituted one great port which had to be governed by the closest co-operative work of all the harbor masters concerned. This was largely due to the unbalanced import trade of London, which so greatly exceeds its export trade that ships leaving their cargoes at London have to pick up cargoes elsewhere.

I made one or two visits to London and to my friends there before Margaret, Barbara, and I went up to Cambridge. We stayed at a delightful hotel near the river and immediately began to inquire of my friends to find out the proper way to get ourselves housed for the winter. Finally, with the aid of an estate agent, we were able to rent a typical English middle-class cottage in New Chesterton, to the northeast of Cambridge proper and not more than a block from where the housing developments began to peter out into open country and farms.

We found a nursery school for Barbara, and thus we were to some extent free from the burden which had been heavy on my wife on the ship—nursing a very sick child and at the same time keeping a very well and lively child in some sort of discipline. Peggy had recovered from her acute illness but was still in delicate health. We secured a Yorkshire woman who had worked at our hotel to take care of the house and hired a car for the long cross-country journey to Tilbury and back.

This was my first auto trip of any length in England, and I

marveled at the abrupt turns of the narrow roads and the way that the main street of each village seemed unfailingly to be on the route taken by our car.

In a few weeks Peggy had recovered most of her strength. There was, however, a stubborn low-grade infection of the middle ear with which we still had to contend, and indeed it was years before Peggy again showed the full vigor which she had had before the trip. Now that they are both grown women, it is still a little surprising to us to see that Peggy is the really vigorous one, having thrown off all her youthful difficulties, while Barbara is somewhat slighter and frailer.

The Yorkshire nurse soon gave Peggy the native accent, no trace of which, happily, remains, though Peggy used to say to me, "Ah coom to Daddy," in the best manner of the West Riding.

Our house had a long narrow garden, with a garden house at the back that could be turned to suit the direction of the sun and the rain. The very friendly lady who rented us the furnished house, and who lived only a door or two from us, secured us the services of a gardener.

Not only was this lady most cordial, but the neighbors in general began to drop in on us from the very first day, with a quicker and readier cordiality than that which I had found in any Boston suburb. This certainly belied the legend of British unapproachability, but perhaps our peculiar position in the community had something to do with it. Most of our neighbors were university people. Not only was I a university man, but I had been at Cambridge many years before and had a certain connection with the place. As Jessie Whitehead has said to me, I must wear the arms of Trinity College with the difference of a bar sinister. Yet, although I am a Cambridge man only on the left hand, the Cambridge family has at least acknowledged me. At any rate, the reader in Italian, who lived two doors from us, and the lecturer in tropical agriculture, who lived next door, visited us at the earliest possible moment and had us over repeatedly to informal teas. This was precisely the easy hospitality which was most to our taste.

Later on we made the acquaintance of the reader in Hebrew, who lived a few hundred yards down the street, and who was official rabbi simultaneously for Oxford, Cambridge, and one of His Majesty's jails. He was a fellow of Queen's College, where he had grown up together with a covey of High Church parsons. He was indeed very much the Jewish equivalent of a High Church parson. He was a ritualist to the last degree, and the door posts of his house had the traditional Hebrew texts fastened to them.

Like so many High Church parsons, he was extremely liberal in his opinions. When it was necessary to raise funds for the repair of the old Norman round church, it was on his lawn that the garden party was held. He had a most interesting collection of seventeenth- and eighteenth-century prints of synagogues and Jewish ceremonies in Holland, among both the Ashkenazic and the Sephardic groups. His particular hobby was to find some of the sources of the Gregorian chants and of Christian ritual in Jewish music and rites.

During my stay in Cambridge, I would run down to London to try to sell the English rights to the filter inventions which Lee and I had made. However, it was no go, largely because a communication engineering did not then seem to be a truly indigenous British industry but rather the reflection of more active interests in the United States and Germany. Furthermore, it became clear to me on a little investigation that a British patent—or, for that matter, almost any sort of European patent—was not something to be taken out as lightheartedly as an American patent (which is in effect nothing but a license to litigation) but involves a considerably greater amount of research and conveys a much greater expectation that the courts will uphold it. It guarantees greater rights and is proportionally more expensive to obtain. In most countries, the laws are less favorable to paper patents than they are in the United States and make very explicit demands that a valid patent be exploited in industry in order for it to remain alive.

While I was in London I met Miss Cartwright, who has since become Mistress of Girton College. She was then, as she is now, a delightfully sincere and unpretentious mathematician in the first rank of English mathematicians, men or women. She invited me to tea to meet a young Trinity don named Paley, of whom I was to see much in the future.

Soon after I arrived in Cambridge, I made the acquaintance of the Youngs, brother and sister. Young was then a don at Peterhouse, I believe, and had done some interesting work which we continued to develop together. Miss Young, a Girton don, was particularly kind to my daughters and she invited all four of us to tea in her rooms.

But of all my scientific contacts in Cambridge, the most important remained my teacher Hardy and his other self, Professor Littlewood. By now, Hardy had become an aged and shriveled replica of the young man whom I had met in Russell's rooms in my student days. Yet he continued to be a redoubtable tennis player and an enthusiast for cricket who knew every finesse of the game. Later on, during his many visits to the United States, he developed an almost equal interest in baseball, and Babe Ruth became a name as familiar in his mouth as that of the cricketer Hobbs.

Littlewood was near the peak of his rock-climbing career. He used to invite me to his rooms at Trinity and demonstrate some interesting rock climbing maneuvers by taking traverses about the base of the pillars in Neville's Court. Hardy and Littlewood took me to see a cricket match. They also showed me a rugger game, in which the players on the two sides managed to pile themselves into unbelievable heaps after a scrummage. I am afraid I did not understand the finer points of either game.

Hardy was lecturing on the elementary theory of numbers, but I did not follow any of Littlewood's lectures. However, I did go occasionally to the mathematical seminar which he held in his rooms.

Unlike Hardy, who had a hatred for the applications of mathematics, particularly those to engineering and warfare, Littlewood has a very considerable physical sense, and played an important part among the mathematical advisers to the military of both the First and Second World Wars. In the First World War he discovered a very brilliant way of computing trajectories of missiles and in particular of making use of a few computed trajectories to obtain a whole range table by interpolation. In the Second World War he and Miss Cartwright left their own chosen field of abstract mathematics for a very useful study of differential equations.

I was invited almost every week to dine at high table either at Trinity or at some other college, and indeed Trinity treated me almost like a supernumerary don. I found the table talk most interesting and less a formal game of high-pressure wit than I had been led to expect. Indeed, when I later visited my academic friends at Oxford, it became clear to me that there was a great gulf between the two schools and that the high-pressure wit which had not been too greatly in evidence in Cambridge was in fact the order of the day at Oxford.

One of the delightful experiences which I had at Cambridge was to go out after the port and a cigar in the Trinity common room following high table and participate with the dons in a game of bowls. I was as poor at this as in all sports of skill, but it was a pleasure to relax with my friends in this cozy and almost secret little garden as the long English dusk came on. It was bowls and not bowling for these two sports must be carefully distinguished. The English sport, which dates back far beyond the time at which Drake played it on the Hoe at Plymouth while waiting for the Spanish Armada, has nothing to do with bowling alleys or tenpins. It is a much closer relative of curling, which, in fact, is nothing but a variety of bowls played on the ice.

The make-up both of the undergraduate body and of the dons at the two universities had greatly changed since the war. No longer were universities the privileged property of the ruling classes. They were now full of brilliant young men who would not have been there at all if they had not been subsidized. Indeed, most of the honors men were subsidized, and the undergraduate who came up for a pass degree was no longer looked on with favor.

Some of these subsidized students came from very poor families and bore the marks of earlier poverty and malnutrition in their stunted forms and poor teeth. Many of them, in addition, had a social hurdle to overcome, which in most cases was more in their own consciousness than in the minds of those about them. I have had young friends from the two universities—and especially from Oxford—tell me how carefully they worked to learn to play the conversational game at the high table. The background of this problem will be clear to all who know D. H. Lawrence and his work, for on the literary side he represents the precise analogue of many of the young mathematicians whom I met in England and, later, as Commonwealth Fellows in America.

The year I was at Cambridge was a great one in the annals of physics, for Cockcroft and Walton had just split the atom for the first time. I saw their apparatus—a pile of glass cylinders and glass plates with holes cut through them, joined together with De Kautinsky cement, which is a glorified form of low-vapor-pressure sealing wax. I was impressed then and I am impressed now with the way in which the English physicists, and indeed European physicists in general, instead of waiting (as do so many American scientists of the present day) for an enormous appropriation, use the material at their disposal and make ingenuity do what one might have thought only money could accomplish.

Yet there was one laboratory at Cambridge which involved great expenditures. This was the magnetic laboratory of the Russian physicist, P. L. Kapitza, who designed powerful generators to be short-circuited and thus put enormous currents through great leads which threshed around like angry snakes under the tremendous magnetic fields thus created. Later on Kapitza was to go back on a visit to Soviet Russia, which has kept a tight hold on him ever since—whether with or without his consent no one will know. At any rate, the Russians sent to England for the whole equipment of the laboratory. He became the pioneer in Russia of that large-scale, factorylike type of laboratory which had first been employed by Kammerlingh Onnes in the Netherlands for low temperature research, and which is now the standard means of exploring the nucleus and of designing atomic bombs. Thus, I felt sure, as soon as I heard of the atomic bomb and of our use of it, that with Kapitza to train the Russians in the technique of this sort of laboratory it would not be many years before they would have mastered for themselves the principles and technique of nuclear research, whether or not they might capture our secrets by means of espionage or persuade a group of malcontents to serve their purposes.

I found a couple of American friends at Cambridge. One of them was a young lady who had done a thesis under me on coherency matrices and was taking a year off from her academic job to combine research with leisure at Cambridge. The other had been an instructor in the Mathematics Department the second year I was at M.I.T. Later on he had gone to study at Munich, where he had managed to get himself embroiled in a challenge to a duel with an army officer while helping an American lady onto a streetcar. He escaped the duel by the ingenious device of using his right to name the weapons. He proposed that these should be bows and arrows and then had his friends let it out that he was an archer of great prowess. Some of us at Tech had seen the story in the Paris edition of the New York Times and had written to him a letter purporting to come from some archery association and offering him the highest honor it had the capacity to bestow. He had always been a bit gullible, and he swallowed the bait hook, line, and sinker. But we could not help admiring the genuine courage of the man, and I think his honest and sincere letter turned the laugh rather against us. At any rate, we found him in Cambridge as a great deal of a companion and a little of a butt.

I tried to walk my young American friend down on a hiking expedition around Cambridge. This time, though, the joke was on me, for he walked me down instead. In fact, he spread the tale of my injudicious challenge to his ability so far and wide that later, when another American went for a hike with me in the Lake Country, I was hard put to it to keep up with his challenge.

As the autumn term drew to a close, Professor Hardy broached two matters to me. One concerned the possibility of my having a book on the Fourier integral accepted by the Cambridge University Press, another the curious Cambridge custom by which a professor may give his lectures by deputy. Professor Hardy had the right to authorize any person to give a course of lectures on his behalf, and these lectures were to be regarded by his College, the University, and the Board of Studies exactly as if he had given them himself. Thus, I was to be a quasi don at Cambridge and to lecture during the second term on my own work on the Fourier integral, even as I had many years before been a quasi undergraduate and attended courses without immatriculation, on the basis of my Harvard connections.

As the term drew to an end and the Christmas holidays came near, I began to receive invitations from my Continental colleagues to lecture at their universities, some of which were still in session. Professor Wilhelm Blaschke invited me to speak at Hamburg; Professor Karl Menger, of Vienna, offered to put me up for a couple of weeks in his rooms; and Professor Philipp Frank, of the German University at Prague, asked me to come there for a few lectures.

Merely to think of these names brings to my mind the vicissitudes which these men have suffered since. Blaschke became, during the Second World War, if not an ardent Nazi, at least an ardent supporter of the Nazis, and he wrote articles in ridicule of American mathematics. In particular, he showed his contempt for the mathematical school of Princeton, which he called "a little Negro village." Menger later on came to the United States as a refugee. I helped him find a position at Notre Dame. I believe he is now in the Illinois Institute of Technology. Frank also came over as a refugee from Hitler and has recently retired from Harvard.

Many of the other mathematicians whom we met on this Continental trip—in fact, most of them—are in the United States or have died since. Hahn, of Vienna, is dead; Artin, of Hamburg, is a professor at Princeton; Gödel, who was Menger's assistant in Vienna, is also at Princeton, where he has done much of his great work on the basis of mathematical logic. Von Mises, of Berlin, whom I met later on the trip, was another Harvard professor until his recent death. In fact, the mathematical schools that were then dominating the Continent have either been transferred bodily across the seas or have gone out of existence completely, and only a very small proportion of the younger men remain to take on the desperate task of rebuilding.

We had a delightful time at Hamburg, where we were taken into the life of the Mathematical Institute. Margaret, the children, and I then entrained for Berlin. There Margaret left me and took the children to see her kinsfolk in Breslau while I continued alone to Vienna.

I had earlier decided to give a few lectures at Prague, and in the course of my travels I wrote to President Masaryk, my father's old friend, hoping to visit him while I was in his country. I therefore took the liberty of writing to him, identifying myself as the small boy whom he had so often seen when he visited my father's house and saying that I was soon to be in Prague. I received a prompt answer inviting Margaret and me to call on him in his palace at Lana. We crossed the Giant Mountains at exactly the best skiing season, and I wished as we passed through that we might visit them in some future winter.

Our friends met us in Prague, saw to it that we found a suitable hotel, and extended to us every cordial hospitality. I was much touched when some of the faculty of the Czech university forgot their traditional feud with the faculty of the German university and came to my lecture.

Finally the day came when we were to drive out to Lana. A state automobile called for us at our hotel and drove us to the palace, through some very bad roads across what seemed quite prosperous farming country. There Masaryk's daughter was waiting for us in a cheerful room with a tall Christmas tree and a blazing fire, until her father should come back from the horseback ride that was part of his daily regimen. It was clear from the informality of our reception that we were being received not as guests of state but as family friends.

The hearty, bearded, old President came in in his riding clothes. He remembered very well his visits to our house on Medford Hillside and cautioned me that I was getting too fat and that I should take more exercise; that I should go in for horseback riding like himself. He expressed great worry at the advances of the Nazis and showed very little hope for the future of Europe. Then he left to take his rest; and after a few minutes' more conversation with his half-American daughter, we left too.

Margaret soon returned to Breslau, and I took the train to Leipzig for a brief visit to Cousin Leon and a talk or two at the University. I think that this was the time I met Koebe, a ponderous, pompous man—"the great expert in the theory of functions," as he was said to have been called by the passers-by in his native town in Brandenburg. There were many tales current about him. On one occasion, when he had visited Da Vinci's horribly mutilated painting of the Last Supper, he is supposed to have said: "How sad! This painting will pass away, while my theorem concerning the uniformization of analytic functions will endure forever!"

Shortly thereafter, we took the train for Holland and the cross-channel boat. By now the children were quite well, but we were exhausted. Luckily, they slept through the crossing, which was as rough as only a winter crossing of the North Sea can be, but we two parents spent the night groveling on the floor of the cabin in an agony of seasickness.

We arrived in Cambridge the next day to find that the frost had burst the plumbing in our bathroom, and we went through two days of annoyance until the plumbers could repair it. However, during this time I had to get my course started, Barbara had to go back to school, and we had to make the best of things.

My course and my book went on very well, and the term continued to follow the pattern of the previous one. I did a large part of my reading in the Philosophical Library, the library of the Cambridge Philosophical Society, in the publications of which some of my earliest work had appeared. It was in this library that I made one of my closest friendships.

I used to read a lot of popular stuff for recreation: particularly detective stories, and the English popular periodicals such as *The Strand* and *Pearson's*. One day I saw in *The Strand* a first-rate thriller called "The Gold-Makers." It was science fiction, with some very plausible science and economics in it, and it had an excellent plot, with conspiracy, pursuit, and escape. It was written by Professor J. B. S. Haldane, of Trinity College, Cambridge. There on the cover stood the photograph of a tall, powerfully built, beetle-browed man, whom I had repeatedly seen in the Philosophical Library.

The next time I saw Haldane at the library, I summoned up my courage to speak to him, to introduce myself, and to express my appreciation of his story. There was one little point in his story, however, which I called to his attention. He had used a Danish name for a character supposed to be an Icelander.

Haldane welcomed this impertinent suggestion of mine, and in a few weeks he invited us to his charming house in Old Chesterton. He first appeared at our house, however, during the Easter vacation, when I was away with an American friend on a hiking-plus-bus trip to the Lake Country. Margaret had never seen Haldane, and he was a little too shy to make clear who he was. However, she recognized him from my description. Haldane invited Margaret to his house to meet Mrs. Haldane. She happened to have gone to London by car and was rather late in returning through the traffic. Hour by hour went on, and still no Mrs. Haldane. Haldane began to be embarrassed, but Margaret was not, and at a late hour Charlotte Haldane turned up. She was very grateful to Margaret for taking the situation with *savoir-faire*.

She was a brilliant young Jewish journalist and novelist, and very charming to Margaret. It was agreed that as soon as I was back from the North, the two families would have a get-together.

Meanwhile, I was hiking over a region which both reminded me of my beloved New Hampshire and contrasted with it. Windermere made me think of Winnipesaukee, but is narrower and less irregular in shape; and the hills behind it are lower, more barren, and more rugged than those behind our New England lakes. There are groves and woods; but the general picture is one of woods surrounded by moor and farmland rather than one of clearings surrounded by woods. There was snow on the higher ground—not as much as there would be in New Hampshire so early in April but still enough to give an atmosphere of bleakness. The houses are of rough stone instead of the wood to which I had been accustomed; the stone walls between the fields are higher and a bit neater than those at home; and the weather, when I went up Scafell, for instance, was cold, wet, and stormy.

When we came back from the trip, much refreshed, Margaret took me over to visit the Haldanes. I remember that we played a lot of bridge with them—family against family, men against women, or Jews against Gentiles. We also had a chance to converse a good deal, and I have never met a man with better conversation or more varied knowledge than J. B. S. Haldane.

It was not long after I came back from my hike in the Lake Country than I began one day to feel rather ill. The doctor came over and diagnosed my case as scarlet fever. We were terribly grieved to have exposed our friends, the Bisonettes, with whom we had spent the previous day, to this infection, and particularly their young boy. However, there was nothing to be done about it, and the ambulance came to take me to the Cambridge Hospital for Contagious Diseases on the outskirts of the city.

I had a pleasant room opening onto a veranda and warmed by a stove, which was necessary as the weather was still occasionally raw and inclement, though it was well along in May.

The partitions between the different rooms were of glass, and as soon as I began to be on the mend, which was in a very few days, I found the chance of playing a non-contagious game of five-in-a-row with the occupant of the next room on a piece of paper pressed against this glass partition. I also had plenty of visitors from the university—my friend Paley came particularly often—and it was not impossible for me to do a little work and to correct a little proof. What I missed most was that I could not participate in the jollities and gaieties of the May Week, which came early in June (when I was already a convalescent), but some of the nurses furnished me with the complete gossip and scandals of the town.

When I got out of the hospital, Term was already over and life in Cambridge had the usual dull flavor of life in any academic town during the summer vacation. However, we continued to see a lot of the Haldanes, and I used to go swimming with him in a stretch of the River Cam, which passed by his lawn. Haldane used to take his pipe in swimming. Following his example, I smoked a cigar and, as has always been my habit, wore my glasses. We must have appeared to boaters on the river like a couple of great water animals, a long and a short walrus, let us say, bobbing up and down in the stream.

Later in the summer I was to participate in the Zurich Mathematical Congress, so Margaret, the girls, and I left for Switzerland. We returned to the Hotel Belle-Rive at Bönigen. Our good friends, the proprietors, were still running the place, although it was about to fold up.

We found a local farm girl who agreed to come with us to Zurich as nursemaid, to take care of the children while we were busy at the meeting. Even while we were on the train to Lucerne and eventually to Zurich, we found the clan of mathematicians beginning to gather. Our old friend Emmy Noether—probably the best woman mathematician there has ever been—whom we had known in our Göttingen days, was on the train, looking as always like an energetic and very near-sighted washerwoman. She was, however, a very warm personality, and her many students flocked around her like a clutch of ducklings about a kind, motherly hen.

We took an inexpensive lodging at a Christliches Hospiz on the hills behind Zurich. This was the Swiss equivalent of a YWCA or YMCA hotel. The place was a bit sanctimonious, but the meals were good, the surroundings delightful, and best of all there was a little zoo in the neighborhood where the children could amuse themselves among a collection primarily of baby animals of all sorts.

As is usual at such congresses, we had a very active scientific and social life, with excursions and entertainments, public and private, at which the university and the Federal Institute of Technology put themselves out in a pleasant competition to do their best by us. The place was full of our old friends, as well as of those who were to become our new friends. By this time I had acquired sufficient prestige to be asked to preside at one of the polyglot section meetings. It was not an easy problem to judge a dispute between a quarrelsome Italian speaking very bad French and an equally quarrelsome German with almost no French at all.

On one of the excursions—I believe it was a steamer trip up the lake—a couple of the Italian mathematicians broached to me the subject of an invitation to lecture in Italy. I had no sympathy with Fascism and resented the completely Fascist and official auspices of this invitation. I talked the matter over with Leon Lichtenstein, who was also a participant in the meeting, and he told me to forget the politics and accept the offer. However, I heard nothing more of the offer, and they must have come to the conclusion that my opinions would not go well in Fascist Italy.

Paley was at the meeting, and he told me that he had made arrangements to come to the United States in the fall and work with me. I was a little alarmed, however, at the way in which he radiated British superiority and railed at the shortcomings of the unfortunate Swiss. I took this opportunity to take him down a peg or two, for I felt that it was easier to criticize his excessive boyish nationalism here, on ground, where we were both foreigners, than it would be later in the United States, where I should owe him the courtesy of a host.

At the end of the congress, tired and let down, we took the train across Germany to Hamburg, where we embarked in a North German Lloyd boat. We were eager to get back from what had been a holiday, if not a vacation, and to resume a reasonably routine life.

8

Back Home.

1932—1933

We came back to Boston well and happy. Margaret busied herself in finding a new house, but even before we had a chance to get settled there, I had a hurried call from my parents. I was used to these unexplained summonses, but I frankly could not see anything in my recent conduct to account for the emergency nature of the call and the tension that my parents' voices expressed.

I found them in a high dudgeon. Father had received a very insulting letter from a philologist colleague in Germany with whom he had tried to correspond, and both my parents considered that my contacts with German mathematicians, who had not the slightest connection with this philologist, must be interpreted as disloyalty to the family.

My father's feelings for Germany were, of course, ambivalent. He had come to hate the tradition in which he grew up and felt personally rebuffed by the very sources from which he would most have welcomed approval. With my mother, the feeling was partly a transference of my father's emotions and partly a desire to assert her solidarity with the prevailing opinions of the Harvard academic group and thus to underline her ultra-Americanism.

Margaret and I had no real idea of what we might have done to merit the stream of invective which bore down upon us. Morever, it was not the usual more or less good-natured invective to which we had been accustomed; it had in it a certain ominous stridency which suggested that Father was in a critical state. Something was wrong, something more than a mere ordinary fit of anger.

My parents went out for the evening, leaving us alone in the house. They soon returned in a state of deep alarm. We learned that Father, while crossing the street, had been knocked down by an automobile. He did not seem to be injured seriously, and although one leg was put out of action, it appeared to be more the pain of the bruise that bothered him than any deeper injury. We called the family doctor—an elderly gentleman who belonged to my parents' generation and had the courtly manners which meant so much more to them than the brusque efficiency of the doctors of my own generation. He decided to leave Father at home for the night.

The next day found Father no better, and some twenty-four hours after the accident he was sent to the X-ray room of the Mount Auburn Hospital. There was a fracture of the neck of the femur, and we all knew that Father was in for a bad time. Before much of a surgical nature could be done, it was necessary to allay the pain and to calm his excitement and forebodings of doom.

The drug chosen was paraldehyde, which is generally one of the more innocent sedatives, but in Father's case it proved to have most undesirable reactions. Much of the time he was in delirium. Excited patients are not welcomed in general hospitals, and it was necessary to send my father to a special institution for cases of this sort. By this time, however, a surgeon had set the hip.

Thus, at the very beginning of my return, I was faced by the need of daily visits to my father at the hospital and of further visits to a second hospital to take him out in our car for an airing in the country. Gradually his hip healed and his state of confusion faded away. I took him back to his apartment, from which he ultimately emerged each day to continue his researches at the Harvard Library, but the almost boyish exuberance which had characterized him before the accident was gone, never to return.

It was during this very trying time that I learned that my English colleague, Paley, was arriving to take up a year of work with me on a Commonwealth Fellowship. I crossed the dismal wastes of East Boston to meet him at the pier. There he was, with a couple of big bags which only as sturdy a man as Paley could carry with ease and with another great bundle of skiing equipment.

The ensuing few weeks were for me an alternation between visits to my father and the most active possible joint scientific work with Paley.

Paley had an unlimited admiration for Littlewood, and I imagine that this had been greatly increased by Littlewood's accomplishments as an Alpinist. It was from Littlewood as well as from his own impetuous, untamable nature that he drew the *élan* which enabled him to drive through any problem that he could not turn. He was the leader of the young generation of British mathematicians, and if he had not come to an untimely end he would be the mainstay of British mathematics at the present moment.

Paley and I used to work together on the great blackboard

of a dusty, half-lighted, abandoned classroom which had been turned into the lumber room of the M.I.T. Mathematics Department. We decided to continue some work that I had already been doing on the design of electric circuits, and we attacked the problem with every tool in our joint repertory. My role was primarily that of suggesting problems and the broad lines on which they might be attacked, and it was generally left to Paley to draw the strings tight.

He brought me a superb mastery of mathematics as a game and a vast number of tricks that added up to an armament by which almost any problem could be attacked, yet he had almost no sense of the orientation of mathematics among the other sciences. In many problems which we undertook, I saw, as was my habit, a physical and even an engineering application, and my sense of this often determined the images I formed and the tools by which I sought to solve my problems. Paley was eager to learn my ways, as I was eager to learn his, but my applied point of view did not come easily to him, nor I think, did he regard it as fully sportsmanlike. I must have shocked him and my other English friends by my willingness to shoot a mathematical fox if I could not follow it with the hunt.

One interesting problem which we attacked together was that of the conditions restricting the Fourier transform of a function vanishing on the half line. This is a sound mathematical problem on its own merits, and Paley attacked it with vigor, but what helped me and did not help Paley was that it is essentially a problem in electrical engineering. It had been known for many years that there is a certain limitation on the sharpness with which an electric wave filter cuts a frequency band off, but the physicists and engineers had been quite unaware of the deep mathematical grounds for these limitations. In solving what was for Paley a beautiful and difficult chess problem, completely contained within itself, I showed at the same time that the limitations under which the electrical engineers were working were precisely those which prevent the future from influencing the past.

The difference between Paley and me was essentially the same as the difference between a great but traditional English classical scholar and my father. I loved my father, and, for all his severity, I knew his softer sides; but he was no Mr. Chips. What was for a Mr. Chips a subtle and difficult game was for my father a struggle in dead earnest to use one's ideas as tools in the world of life outside. I respect and I understand English scholarship, but my roots are Continental.

Paley was a skiing enthusiast but not a particularly skillful skier. He had the habit of going deliberately over forbidden slopes, and he would stamp his heavy weight on his great long skis in a way that would make the local ski master cock his eye in alarm. In fact, Paley's whole policy of life was to live dangerously. For him, any concession to danger and to self-preservation was a confession of weakness which he dared not make in view of his desire for the integrity of a sportsman.

Paley visited us in New Hampshire. He showed a remarkable lack of understanding of the life there and of the way to get things done. After my experience of his tactlessness in Switzerland, I was not surprised at this. When his car gave him some difficulty on the way up to New Hampshire he had tried to pay the passers-by who had helped us to get going again, even though I told him that these were our neighbors, who were helping us out of friendliness, and who would bitterly resent any attempt to pay them off.

Yet for all this, as he came to learn what life was really like in America and in the American countryside, he began to respect our people for their friendliness and independence. The fact is, Paley remained a schoolboy until his death, and he would have remained a schoolboy still if he had lived to be eighty.

We were to learn later that the front of boldness and danger seeking which Paley put on was part of a true death impulse. At any rate two cousins met as violent ends as Paley himself was later to do. One of them died in a motor smash and the other in a rock climbing accident.

During the Christmas vacation, Paley did a little skiing in the Adirondacks with an Irish friend who, I think, was another Commonwealth Fellow, and after the skiing was over they continued to Montreal. I believe they almost wrecked their car in the Adirondacks, and they got entangled with a group of New York gangsters who had moved to Montreal during prohibition. Paley came home to Boston rather thrilled than chastened. At that time I already knew that the best thing that could happen to him would be some really dangerous and scaring experience—some accident short of death.

In April he went with some Boston friends for a skiing party in the Canadian Rockies. There were safe and interesting ski slopes near where the party was encamped, but the ski master had put some neighboring slopes out of bounds as too likely to have avalanches. To forbid a thing to Paley was to make quite sure that he would do it.

I shortly received a telegram from his companions that he had been killed in a ski accident. He had gone across one of these forbidden slopes at an angle. Apparently he had stamped with his heavy skiis at a place where an avalanche was just ready to start. His body was found a day or two later three thousand feet down the mountain, with one leg torn off. He was buried at Banff, and it was my sad duty to inform his mother and his English friends. It took me some time to come back to a mental equilibrium sufficient to permit my further work and my proper attention to the environment about me.

After this, I turned to the group of my friends centering about the Mexican physicist Manuel Sandoval Vallarta. He introduced me to a Mexican physiologist named Arturo Rosenblueth, the right-hand man of the great Harvard physiologist, Walter Cannon, whom I remembered meeting when I was a boy of about eight. They had worked together in many fields of physiology, in particular neurophysiology. By that time, it was already clear that whatever Arturo's academic fate at Harvard might be, it was his scientific destiny to carry on the great Cannon tradition.

Arturo is a burly, vigorous man of middle height, quick in his actions and in his speech, who paces rapidly up and down the room when he is thinking. No one who sees him in the Mexican environment can doubt that he is a true Mexican, though the greater part of his genetic heritage comes from other countries, particularly Hungary.

Arturo and I hit it off well together from the very beginning, though to hit it off well with Arturo means not that one has no disagreements with him, but rather that one enjoys these disagreements. One point that we shared in common was an intense interest in scientific methodology; another, that we believed that the divisions between the sciences were convenient administrative lines for the apportionment of money and effort, which each working scientist should be willing to cross whenever his studies should appear to demand it. Science, we both felt, should be a collaborative effort.

Arturo carried out at least part of his ideas of what science should be by running a private seminar on scientific method at the Harvard Medical School. Not all those who attended were medical men, however. Manuel Vallarta as well as some other people from M.I.T., including myself, became regular attendants. A few from the branches of Harvard on the Cambridge side of the river also frequently sat in on the meetings. Naturally, it was Arturo who was the chief host in these feasts, which belonged specifically to him, but if there was any other person who took the reins into his hands, I think that by general recognition that person was myself. Thus, the seminar represented some years of our preliminary cooperation before the final and definitive co-operation which has taken place within the last twenty years.

While our seminar was at no time a part of the official teaching of the Harvard Medical School or of any other institution, its many alumni will give it the credit of founding in them a wide interest in the philosophy of science and even of initiating them in quite specific lines of thought and research. Since Arturo has left the Harvard Medical School for Mexico, the two of us, either jointly or singly, have continued similar meetings in Mexico City or at M.I.T. We have managed to recreate much of the atmosphere of those earlier gatherings, although it has perhaps not been possible for us to bring back the full flavor of active enthusiasm which belonged to our earlier days, when our main successes were ahead of us. In its later days, the seminar established a scientific reputation which may have tempted certain eager beavers to try to take over the prestige of the group for their own advantage.

Ultimately the theme of the many discussions which Rosenblueth and I had privately and in our seminar came to be the application of mathematics, and in particular of communication theory, to physiological method. We laid out a policy of joint effort in these fields for a future when we might work closer together.

9

Voices
Prophesying
War.
1933—1935

During my stay in the English Cambridge, the depression had reached its height. When we had come back we found that it had entered a new phase, and that the dangerous possibilities already implicit in it were well on the way to being realized. England had devalued gold at about the time Paley arrived, and he assured me that this was a move of extreme cleverness which would ultimately give England a decisive economic advantage. He told me quite correctly that it was a move that we should be forced to follow ourselves, but that the first nation to act gained an advantage thereby that could never be made up by those who followed suit. It was clear that the world was going to be subjected to shocks which the patched-up economic system that came into being after the First World War, and the patched-up social system that went with it, might not be able to sustain.

When I had at last been in contact with Leon Lichtenstein, at Zurich, I had found him in a depressed mood, owing to the political ground that had been gained by Adolf Hitler and the Nazi party in Germany. He knew that the *Putsch* meant trouble, and trouble was not long in coming. We read in the papers of the anti-Jewish measures, and even as these appeared in the foreign news bulletins Leon sent us a pathetic and desperate letter. He had not waited for the *Putsch* to flee Germany for Poland. He wrote me from a hotel in Zakopane in the High Tatra to find a job for him in the States.

I set out at once to look over the ground, but before I could get started we received another communication—I think from Lichtenstein's wife—that Leon was dead of heart failure. Then I knew that the work of us American mathematicians was cut out for us and that we should have to get together and make a systematic effort to find jobs and a possibility of life for many a displaced scholar.

About this time I received a very tragic letter from Mrs. Szász, the wife of a lovable little Hungarian who had taken me under his wing in my student days in Göttingen. Luckily there was a Mathematical Society meeting not far off, and I had a chance to talk with a Cincinnati colleague, Irving Barnett, the very same man who had introduced me to the theory of generalized integration. Barnett offered to place Szász in Cincinnati in the future, but the Massachusetts Institute of Technology was ready to accept him for a year or two until he could become more proficient in the English language.

Soon Szász turned up, and later he brought his daughter and his brother over to this country. He had a career of some twenty years in America, where he ultimately received recognition more appropriate to his really very considerable talents than he had found in Germany.

Szász was but the first of a great stream of *émigrés,* some of whom went through my hands. I had at least some share in finding positions for Rademacher, Pólya, Szegö, and many others.

Rademacher, of Breslau, was invited to the University of Pennsylvania by Professor J. R. Kline. His field in the analytic theory of numbers was one which had not been too strongly represented in the United States before his arrival; but he formed a school here, and much of his best work has been done in the United States.

Pólya and Szegö are two Hungarians who are particularly strong in the classical field of analysis. Between them they had already written a very interesting sort of textbook which consists in a large repertory of research problems at just about the level at which the novice in research can undertake them with profit. Both of them were received by Leland Stanford University, and Pólya has just retired there.

Emmy Noether, the greatest woman mathematician whom the world has ever seen, was cordially received by Bryn Mawr. Unfortunately, after all too few years she died there from causes presumably unconnected with the Nazi regime.

Menger wrote to me from Vienna to see if he could find a refuge in the States before the Austrian crash came, and we secured him an invitation from Notre Dame University. He has since moved on to the Illinois Institute of Technology.

Von Neumann, who was something of an infant prodigy, received a cordial welcome at Princeton, together with Einstein and Hermann Weyl—who is without doubt the greatest German mathematician since Hilbert. They all later moved to the new Institute for Advanced Study, founded at Princeton with the advice and under the administration of Veblen. Veblen and Kline were perhaps the two chief American almoners for the European refugees, but I am proud to believe that I did at least my share in the matter.

From the very beginning of the problem, I tried to get in contact with Jewish charities and individual Jewish sources of funds for the tremendous task of salvaging as much as possible out of the Nazi mess. Here I found a mixed reception. The Jewish sources of charity had very frequently decided that Jewish displaced scholars were in most cases too far from Judaism to be their special responsibility. Furthermore, it was the time of the height of the Zionist movement, and the Zionists considered that part of their charitable bounty to be spent abroad or on foreigners should go in the first instance to Zionist undertakings, and only in the second instance—if there was a second instance—to other causes.

Even so, we found a considerable amount of Jewish money at our disposal, but I think that at least half of the funds we needed came directly from our already overloaded universities, and from our large scientific fund sources such as the Rockefeller Foundation and the Guggenheim Foundation.

It was about the time of the *émigré* influx that I received an invitation to join the National Academy of Sciences. This is the organization which was entrusted during the Civil War with the task of putting the services of the scientists of the United States at the command of the American government. In the course of the years, its governmental importance had gradually given way to the secondary function of naming those American scientists who might be considered to have arrived. There has always been a great deal of internal politics about science, and this has been distasteful to me. The building of the National Academy was for me a fit symbol of smug pretentiousness, of scholarship in shapely frock coat and striped trousers. After a brief period, during which my inquisitiveness concerning the nature of the high brass of science was amply satisfied, I got out.

It was during the period after my visit to England and before my visit to China that I was perhaps the most active as a participant in Mathematical Society meetings. The very young man cannot afford to go to too many of these. The older man, as I now regretfully find myself to be, does not have the energy to go through too much of the excitement of the meetings and cannot, in fact, attend them without finding claims made on him to act as a presiding officer or to take some part or other in the councils of the group. Moreover, his prestige enables him to let others come to him without the necessity of his going out to meet them. It is therefore the middle period of a man's life which is the most suitable for extensive meeting going.

Among the most delightful meetings were those that usually took place during the spring holidays in the South. To leave half-wintry Boston and cross the cold stretches of Massachusetts and Connecticut, where the merest haze of green foliage was beginning to decorate the bare trees, and to meet in Pennsylvania and Maryland the white of the dogwood and the purple of the Judas tree, was a pleasure indeed. We generally took one of the children along with us, and I think they enjoyed the trips as much as we did.

It was not at one of these delightful spring meetings, but on a bitter Cambridge winter day when the temperature went down to eighteen below and the inadequate steam hissed futilely into the chilly Radcliffe classroom where we held our meeting, that Professor Morse, of the Institute for Advanced Study, and I received the Bôcher prize in analysis together. It was a pleasant thing to be recognized, and even more pleasant to receive this recognition from the mathematician whose opinion I valued above that of all others: Hermann Weyl.

One of the honors that normally comes to the successful American mathematician is a request to write a book for the so-called Colloquium Series. As far as the personal satisfaction of this invitation goes, it is quite equivalent to the award of a prize. The possibility of an invitation of this sort had been hovering about me for some years, and Tamarkin was my most enthusiastic supporter. The invitation came after Paley's death.

If he had lived, we would have divided our joint efforts to the writing of this book; and now that he was dead, I felt it to be a work of piety to incorporate into one book the various pieces of research we had done together and the other investigations I had made along similar lines.

The title of the Colloquium was *Fourier Transforms in the Complex Domain*, and it appeared with the two of us as joint authors. I presented the material of the book at the summer meeting of 1933, which was held in the very delightful college town of Williamstown.

At about this time in the academic year of 1933–34 an interesting group of young undergraduates was coming up in the mathematics department at M.I.T. Claude E. Shannon was a very striking member of this group. He hit on an idea which even then showed a profound originality and which has since been justified by the vast amount of work on switching devices, computing machines, and information theory to which it has led.

Let me point out the content of this idea, whose implications are of major importance in current scientific work. All of you are familiar with the wall switches which turn on and off the lights in a room. In their simplest application, each switch operates one light alone. However, more complicated arrangements occur in almost every house. For example, you can turn the hall light on from the foot of the stairs and turn it out from the head. It is possible to find arrangements by which the same light can be turned on and off from four or five different wall switches. It was Shannon's discovery that the mode of designing these switching circuits with greatest economy is in fact a branch of the algebra of logic.

Switching circuits lie at the heart of automatic telephone central exchanges, and Shannon's special talents proved to be exactly what the Bell Telephone Laboratories needed. Shannon has gone on from triumph to triumph as an employee of those laboratories. His interests have come to embrace the general measure of information, the electrical mouse, which can learn its way around a maze; the automatic chess player; the problem of coding and decoding messages; and in fact the entire scope of modern information theory. In all of this he has been true to his first intellectual love for problems of a sharp yes-and-no nature, such as those of the wall switch, in preference to problems which seem to suggest the continuous or roughly continuous flow of electricity.

Thus, Shannon is one of the major spirits behind the present age of the electronic computer and the automatic factory. Moreover, it is through his work that a training in symbolic logic, that most formal of all disciplines, has come to be one recognized mode of introduction into the great complex of scientific work of the Bell Telephone Laboratories.

While Shannon was an M.I.T. man, and while Bush was among the first of our staff to understand him and to value him, Shannon and I had relatively little contact during his stay here as a student. Since then, the two of us have developed along parallel if different directions, and our scientific relations have greatly broadened and deepened.

Two younger men who made their appearance in our department during the early thirties were W. T. Martin and Robert Cameron. Martin left us to head the mathematics department at Syracuse, and his splendid work in building up that department led to our calling him back as head of the M.I.T. mathematics department. His sincerity and understanding friendship with the other members of the staff have been an enormous asset in bringing our department up to its present status. He and Cameron did a good deal of work together along the lines of my Brownian-motion papers, and they organized the field into a generally recognized branch of mathematical work. Cameron has since left us for the University of Minnesota, but for many years he has continued to work together from time to time with Martin.

There were three young men studying mathematics at that time who became more definitely attached to my line of work. These were Norman Levinson, Henry Malin, and Samuel Saslaw. They were Jewish boys who had felt, in differing degrees, the frustration of prejudice without being so deeply wounded by it as to be completely ruined.

Saslaw was the oldest of the three and acted as a good-natured older brother to the group. Levinson was decidedly the strongest mathematician, and he has been one of my successors as a recipient of the Bôcher prize. He is now at M.I.T. and he is a mainstay of the middle generation of research men. Malin filled a very useful part in encouraging the *esprit de corps* which such a group of young men needs to carry it through the difficult work of mathematical research.

Levinson had done research for me at a very early age, and, even before he had completed his bachelor's work, he had taken a job of mine which was already an extension of Fourier Series theory and pushed it to the extreme limit. I felt that the boy had obtained about everything he could from me at that time and that he needed the broadening of contact with other scholars new to him. Hardy was willing to accept him at Cambridge, and we secured for him a Redfield Proctor scholarship from M.I.T. This was in 1934. Levinson's performance at Cambridge was eminently satisfactory, and it was on the basis of the impression that he left that future American mathematicians in Cambridge have been judged and assessed.

By the time at which these young men had come to work with me, Lee, my Chinese friend, and Ikehara, my friend from Japan, were already gone. I have told in an earlier chapter how we learned of Ikehara's desperate straits and had returned him to his own country.

Lee also had gone back to China, on his own initiative, to seek a job there. He had tried both government and commercial work but had found that in neither was it possible for a man to live up to the demands of his own probity. Luckily he had been appointed a professor of electrical engineering at Tsing Hua University, which was just developing from a receiving institution for those boys destined to be Boxer Indemnity scholars in the United States into a full-fledged and autonomous university. Here he was finally in a congenial environment.

Lee had not forgotten me. In the course of the academic year 1934–35 I received an invitation from Tsing Hua to come there for a year and lecture on mathematics and electrical engineering. The invitation came, of course, from the Tsing Hua authorities, and particularly from President Mei and Dean Ku, who was later to become vice-minister of education in China. But the invitation had certainly been inspired by Lee. After a fair amount of negotiation, during which I was always eager to go but had certain qualms about the possibility of going and the stability of affairs in China, I accepted. By then my two daughters were seven and five years old, and we decided to take them along, even though we were somewhat worried about the safety of the trip and its effect on their health.

My prospective trip to China filled me with great enthusiasm. Not only have I always loved travel for its own sake, but my father had brought me up to consider the intellectual world a whole and each country, however exalted its position might be, as a mere province in that world. I had actually been a witness and a participant in the rise of American science from a provincial reflection of that of Europe into a relatively important and autonomous position, and I was sure that what had happened here could happen in any country, or at least in any country which had already shown in action an aptitude for intellectual and cultural innovation. I have never felt the advantage of European culture over any of the great cultures of the Orient as anything more than a temporary episode in history, and I was eager to see these extra-European countries with my own eyes and to observe their modes of life and thought by direct inspection. In this I was thoroughly seconded by my wife, to whom national and racial prejudice have always been as foreign as they have been to me. Even my daughters, small children as they were, had been brought up outside of the usual run of prejudices.

We spent the beginning of the summer as usual in our home in Sandwich, and I participated in hikes with my young friends, who were now beginning to grow up. We took the train from Meredith, our usual jumping-off place, and went north by way of Montreal to Chicago. We left for California, where I had accepted invitations from Stanford University for a few lectures. There we saw our friends the Szegös and the Pólyas, together with other even older acquaintances. We left for Japan by a boat of the Dollar line.

10

China

and

Around the World

On the boat for Japan we found ourselves in the company of a mixed group of returning missionaries, Japanese army officers, and conducted tourists, but with two young children with us we led a very self-centered life on board.

At Yokohama we were met by our friend Ikehara, who was teaching at the University of Osaka, and he served as our mentor for the two weeks we were to spend in Japan.

Rooms were waiting for us at the Imperial Hotel in Tokyo, that fantastic structure of the American architect Frank Lloyd Wright, and Ikehara had made arrangements that a Japanese lady who had been in America should invite our children. Lo and behold! This lady had lived in Boston, where we had known her well, for her daughter had been a friend of Barbara's at kindergarten. Thus, my children got along splendidly with their hostess, and we had no difficulties at all in this respect.

The hotel was excellent and the food very good, but in those days the guests were subject to a continuous surveillance. For all we knew, the American lady who sold dolls and souvenirs might have been under orders to report our every word or attitude to the management and eventually to the police. Perhaps the waiters were a little more than waiters. Still, we had a very good time, and Margaret enjoyed shopping for Japanese souvenirs.

During the next few days, in hot August weather, I met the Tokyo university people and gave a few talks. I found the academic standard at the universities quite high, but I also felt that Tokyo in those days had begun to be affected by the rigidity that so often taints a university which is sure of its position as the first in the land. The Tokyo professors looked slightly down their noses at their associates at the lesser universities.

Ikehara accompanied us to Osaka, which was even muggier than Tokyo. I found the mathematics club at Osaka University very much to my liking. It is from this group that many of the best Japanese mathematicians have come, such as Yoshida and Kakutani, who stand among the best mathematicians anywhere.

We visited the great keep of Osaka. There is no European building more suggestive of strength and military impregnability than this cyclopean structure. The battered-back massive stone walls could even now hold an army at bay. It seemed to us a comment in stone on the old Japan of the *bushi* and *ronin*.

We left Kobe for the China coast by a small Japanese steamer. It was a sail of several days through the beautiful inland sea of Japan and then past the forbidding and bare hills of Korea, to the starboard, and Shantung, to the port.

The food was dull and inferior, and the company more interesting as types than as companions.

After a few days we saw fishing boats which had put out
from the low muddy coast, and soon we docked at the pier
by the railway station at Tangku. We were surprised to see
how much taller the Chinese porters on the shore were than
the run of the Japanese we had met, even though we had seen
a few tall and even burly Japanese here and there.

Soon the Chinese custom house officer came on board, in-
quired courteously for us, and told us that Dr. Lee was waiting
for us on shore. Within a few minutes Lee met us and we
were sitting together in the great, empty, waiting room of the
railway station, which was adorned with pictures of the summer resort at Pehtaiho. Lee told us that he was married, but
he would not tell us from what part of China his wife came.
He immediately started discussing with me both the work
we had already done and our prospects of work to come.

Soon the train came in, and we entered a second-class compartment for Peiping. The train had cars which looked American from the outside, although they followed the European
compartment system. Instead of a water cooler there was a
little stove at one end of the car, where the attendant kept a
pot of tea always on the boil for the passengers. Indeed, we
found that boiling tea is as much a necessity of life in China
as cold water is with us.

We traveled through flat country reminiscent of the corn-
fields of Kansas, where we saw corn as well as the tall Chinese
sorghum. The farm houses were of adobe with somewhat less
curved roofs than we had been led to expect from the pictures
we had seen of southern China. Soon the barren Western
Hills came into sight, and before we knew it we were at the
main railway station of Peiping. A seven-miles taxi trip took
us to the South Compound. This was a large aggregation of
modern bungalows which, however, instead of facing one an-

other along streets of the Western type, all faced to the south and looked into one another's kitchen compounds.

There we found Mrs. Lee, who was not a Chinese lady at all but a tall beautiful Canadian, who had been working in New York for the same firm as Dr. Lee himself, and for whom Dr. Lee had sent once he was established in a permanent position. She was in charge of the Lee household of servants, who all appeared at our house to enable us to get started, and who had a delicious meal ready for us.

It took us a few days to get used to Tsing Hua and our new life. It was a bilingual place, where much of the teaching both in the Western humanities and in science was in English. Although there were some Western professors on the faculty, the majority were Chinese, trained for the most part in the United States but some of them in England, Germany, and France, as well.

It was interesting to see how this foreign training had reacted on the members of the faculty. There was a Chinese lady who had studied in Paris and whose walk looked French even at the distance of a couple of blocks. There was an assertive little German-trained professor—nothing but minor details of complexion distinguished his appearance from that of the perfect Nazi. Many of the professors were as American as my colleagues at home, and there was a tweedy professor of English who had Oxford written all over him, and inside his soul as well.

Professor Hiong, the head of the mathematics department at Tsing Hua, whom I had already met in France, was as much at home in Paris as on the edge of the Western Hills, and his eleven-year-old boy was an artist in the Western style. The boy was able to write with ease a French composition for a *lycée* in Paris or a discussion of Confucian virtue in classical Chinese.

The first things we had to attend to concerned the bread-and-

butter necessities of life. The Lees found for us an establish-
ment of servants. There was an elderly butler, or number-one
boy, who had worked for people at the French Embassy. He
was supposed to know a little French, which consisted mostly
of the word *oui*. There was an amah, or maid; an elderly Mo-
hammedan rickshaw boy to take our children to the American
school at Yenching University; and a cook (or "great expert,"
to translate the Chinese expression). He did not live up to his
name, but he certainly could produce acceptable meals.

One of the most important things to find was a Chinese
teacher to come to our house and instruct us in the language.
Lee found for us a tall, dignified, elderly gentlemen in a long
gown, who knew little English and spoke none with his charges,
but who brought with him a textbook in English with the aid
of which he could supplement his oral lessons. We did not
find progress too slow, and both my wife and I took a great deal
of interest in learning a language so new and unfamiliar to us.
Our teacher proved to be more than a teacher, for he was the
vehicle of all the local gossip both to and from our house. We
always knew if one neighbor had an extra big banquet on for
the night, or if another neighbor had servant troubles, or if the
Christian congregation was to meet that Sunday in the home
of a third.

We sent the children to an elementary school carried on
under the auspices of nearby Yenching University. Before
school had started, they had managed to make friends with the
Ni children, who lived nearby and whose mother was American. The two broods raided the neighbors' vegetable gardens, invaded the privacy of the neighbors' houses, and, in
short, got into all the sorts of trouble into which active children
can get.

The American school at Yenching was a one-room structure
containing all grades from the first up to the high school.
The teachers came mostly from the faculty of Yenching Uni-

versity. All the children except mine were so fluent in Chinese that my own made very little effort to bridge the gap between their attainments and those of their schoolmates.

Occasionally my children would be greeted with abusive cries from the village children, with the result that my daughters learned Chinese abuse with an alacrity with which we could have well dispensed. They hurled back the cries of the village children without understanding what they meant. When my daughter asked the meaning of one of these phrases, the Chinese teacher stood her in a corner.

The school was very good for my children. The informal arrangement made it possible for Peggy to advance far more quickly than in any graded school. She was the only beginner, but before the year was half over she had really started to read.

It was quite early in our stay at Tsing Hua that Mr. and Mrs. Hiong were hosts to the whole mathematics department for a picnic on the grounds of the New Summer Palace. This was a rococo assemblage of passageways and temples. At every turn there were windows of fantastic shape in the form of vases, lutes, hearts, or lozenges. There were long winding walks and gingko trees, and staircase upon staircase running up the hills to little summerhouses.

Mrs. Hiong had prepared a collection of tidbits in Chinese style. The mathematics department servant carried the basket and participated as an equal in the feast. The Hiong boys made sketches of passers-by in which they showed a sense of humor at once Chinese and Western.

Everybody could talk English, and all but ourselves talked Chinese. The atmosphere was one of democracy and free interplay of personality with personality. The servants were neither actively snubbed nor tacitly left out of the conversation.

We had later invitations to the Hiongs' house, which was almost a museum of delightful paintings in the modern Chinese style. There were many panels covered with delicate and deft representations of fish and shrimps and crabs and all the other little animals of the water. The Chinese, like their pupils, the Japanese, show a tenderness and taste which is never cloyed with the sentimentalism that spoils so many Western nature paintings. To the Buddhist, God is not separate from nature but is manifest in all nature. Thus, nature contains God himself as a part of its very being and not as part of a quasi-human personality imposed on the world from without.

To go from Chinese painting to Chinese banquets is a short step, for cookery is quite as much a Chinese art as painting. For me, a vegetarian, there was a wealth of dishes—and indeed an entire school of cookery—especially concocted for the vegetarian Chinese Buddhist monks. Yet even my taste for Chinese food scarcely mitigated for us the feasts of twenty or thirty courses. I am afraid that the cookery we asked for at home shocked the professional sensibilities of our "great expert," for we preferred plebeian food to the dainties which gave him a chance to display his skill.

My lectures at Tsing Hua were in English, which all the students could easily understand. More than one of these students later turned to pure mathematics or to electrical engineering. They are now scattered around the universities of China, and of the United States as well. In my spare hours around the classroom building I used to sip the tea that always was kept prepared by the department servant and play innumerable games of chess or five-in-a-row or go with my colleagues. I never mastered the game of go, however, and would now prove a mere child at Princeton among the experts of Fine and Fuld Halls.

I continued to work with Dr. Lee on problems of electric-circuit design. We attempted to do some laboratory work, which came to nothing because it involved technical problems which we were not then in a position to handle adequately.

What Lee and I had really tried to do was to follow in the footsteps of Bush in making an analogy-computing machine, but to gear it to the high speed of electrical circuits instead of to the much lower one of mechanical shafts and integrators. The principle was sound enough, and in fact has been followed out by other people later. What was lacking in our work was a thorough understanding of the problems of designing an apparatus in which part of the output motion is fed back again to the beginning of the process as a new input. This sort of apparatus we shall know here and later as a feedback mechanism.

Feedback mechanisms were already used by Bush in his computing machines, but they have certain serious dangers of their own. Too intense a feedback will make a machine oscillate so that it cannot come into equilibrium. It is not too difficult to avoid this with the relatively light feedbacks used in the machine of the Bush type, but with the heavier feedbacks of the purely electrical machines this difficulty is hard to overcome. What I should have done was to attack the problem from the beginning and develop on my own initiative a fairly comprehensive theory of feedback mechanisms. I did not do this at the time, and failure was the consequence.

My main work, however, was lecturing on generalized harmonic analysis and on the material Paley and I had included in our book. I also was engaged in new pure-mathematics research on the so-called problem of quasi-analytic functions.

In the last hundred and fifty years, analysis, which represents the modern extension of the infinitesimal calculus, has split into two main parts. On the one hand we have what is known as the series of functions of a complex variable, which is a continuation of the eighteenth-century theory of series proceeding in the powers of a variable such as 1, x, x^2, etc. This theory is particularly applicable to quantities which change smoothly and gradually. At one time it was supposed that all important mathematical quantities changed smoothly and gradually; but toward the end of the eighteenth century, the study of harmonic analysis, which is analysis of vibrating systems, showed that curves pieced together out of parts which had nothing to do with one another were subject to an analysis of their own. This point of view first led to the theory of Fourier series and then to the general branch of study known as the theory of functions of a real variable.

The theory of functions of a real variable and the theory of functions of a complex variable thus represent two separate but related subjects, which do not succeed one another like a freshman and a sophomore course but represent radically different insights into the nature of quantity and of the dependency of quantities upon one another. They have had a great deal of interplay in the course of the last hundred and fifty years. However, it is only recently that it has come to the attention of mathematicians that there are certain intermediate fields of work which share the methodology of both. There are certain curves which are smooth enough so that the whole course of these curves is known from any one part, but which are not smooth enough to be treated by the classical theory of the functions of a complex variable. The study of these curves is known as the theory of quasi-analytic functions. The French school of mathematicians has contributed greatly to this field, and here some of the best work is that of Szolem Mandelbrojt. However, the book by Paley and myself in this field has also led to results which I was pursuing during my stay in China.

I had hoped to meet Mandelbrojt sometime soon—as, for example, at the scientific congress to take place in Oslo in 1936. When I heard that Hadamard, Mandelbrojt's patron, was also coming to Tsing Hua University, I hoped that he could arrange for a meeting between us in France while I was on my way to the congress.

I have already pointed out that my work with Paley had

direct applications to the study of electric circuits. These were related to the same issues which also came up in the theory of quasi-analytic functions. Thus, here again, as so often in my work, the motivation which has led me to the study of a practical problem has also induced me to go into one of the most abstract branches of pure mathematics.

This represents an attitude of mine which is in harmony with the greater part of the work of the eighteenth and nineteenth centuries, and which since then has found its representation in the writings of Hilbert and Poincaré. It is not the prevailing attitude at the present time either in America or elsewhere. The two mathematicians now or recently active in America who have adopted a similar point of view are—and I believe not by coincidence—two of the greatest forces in modern mathematics, namely, Hermann Weyl and John von Neumann.

This mathematical work on quasi-analytic functions was, of course, not the whole of my life in Peiping but was interspersed with periods when my chief interest was to behold the rich and unfamiliar panorama of the life about me.

We used to go to the city quite often by bus or taxi or even rickshaw. Leaving for the city and returning by rickshaw was an interesting process, though one felt ashamed at being pulled along by the strength of another man. At the northwest gate of the city there was a teahouse which was a rickshaw man's exchange. The guild of rickshaw pullers outside the walls was separate from that functioning within the walls, and no man from inside the city was permitted to ply his trade further out than this teahouse. Thus the rickshaw boy who took you to the teahouse in either direction made a bargain with his opposite number from the other guild and was paid by the new man for the work he had already done. Later on the passenger payed the new rickshaw man what was owing to both.

When we first went into the city we were beset by a gang of rickshaw men, who almost tore us to pieces in their eagerness to make us their employers. Acting on Lee's advice, Margaret and I each picked one, and we had no further difficulty in this respect. Our steady rickshaw man was almost always ready for us when we came in and got off the bus, and if he was not there he would have assigned somebody else to take his place. These bargains were generally respected by the whole guild.

When we dined in town, my wife would send what was left on the serving dishes down to the rickshaw man for his meal. My rickshaw man was a Moslem and would not accept non-Moslem food, so I would pay him off with a small extra sum for his lunch. Both our rickshaw men were extremely loyal to us, particularly after Margaret's man had damaged his eye in a fight with another rickshaw man and she had sent him a sum of money to help him take care of himself while convalescing. When we finally left Peiping he gave my children Chinese hats to wear, and my rickshaw man gave us a can of fine tea scented with jasmine.

My rickshaw man was both intelligent and enterprising. He read his newspaper, and later in the year when Professor Hadamard came to Tsing Hua University, my puller asked me whether it was true that Ha-Ta-Ma Hsien-Sheng was as great a mathematician as the newspaper said he was. I have no doubt that part of my rickshaw boy's interest lay in trying to get a permanent job with Hadamard, but at least he was both as enterprising and as well-informed as one would expect a very intelligent cab driver to be in the States.

In town, Margaret would shop at the Clock Shop, the Big Bell Shop, and at many other such places. In these higher-class Chinese shops the service was often excellent and even modern, but the décor and the costumes were thoroughly Chinese. The subordinate clerks wore the cotton gown of their class, and the proprietor and his immediate underlings were dressed in gowns of gray or blue silk damask. The language of the shops was either good English or good Chinese. Pidgin English was almost unknown in Peiping. There was a great deal of snobbery and mutual belittlement between Shanghai, the New York, and Peiping, the Boston of China. The Shanghai ladies were quite as eager to keep in the latest style as the ladies of New York, and their Chinese-cut gowns often showed bold and attractive adaptations of Western fashions. The Peiping ladies, on the other hand, dressed with some of the aristocratic intellectual dowdiness which I have seen on the banks of the Charles.

Peiping was an ancient capital with a long tradition of art and culture. Here and there one would see simple farmers of Manchu blood, the descendants of the courtiers of the Ching dynasty, who spoke the beautiful Chinese of the palace.

The city was a mixture of glamor and squalor. It was intriguing to walk down ill-paved alleys which seemed to lead from one slum to another, but where the vermilion moon-gates often opened into a little gem of a courtyard and garden surrounded by pavilions of taste and beauty.

My friends were both American and Chinese. So far as concerned the serious intellectuals on both sides, the relations between the two groups were excellent. On the other hand, the rag, tag, and bobtail of Englishmen abroad, which Thackeray knew so well in the Paris of the Restoration, were to be found in 1935 throughout the treaty ports of China.

The Japanese were moving in. They had already taken over the administration of most of Pei-ho Province, either directly or by satellites. The town was full of Japanese soldiers whose drill would carry them brutally into the midst of a Chinese crowd.

I went down with Lee to Tientsin to collect part of the pay for our inventions from an agent of the American Telephone and Telegraph Company. Tientsin was interesting. With its several foreign concessions, one would go in one block from Russia to France, and in another from France to England.

There was something unreal in this finely subdivided macé-
doine of different nationalities.

The student life was greatly disorganized. The students put
on a strike that lasted for months. They marched into the city
to protest the coming of the Japanese and the supineness of
the Chinese. When the students were not allowed in at the
railroad gate, a little wisp of a girl rolled under it and opened
it. Inside the city, the police force used clubs and fire hoses on
the students, and the hospitals and jails were full. However,
many of the students came from the leading Peiping families,
so their parents found a way to compromise.

All this time Lee and I spent in work. We would sit over the
drafting board in Lee's den while our wives would talk and
read in the next room. After they thought we had had a sufficient dose of work, they would call us out for a snack and a
cup of tea, and finally we would finish the day with a game
of bridge.

One of our big problems was to get our patent application
through the American Patent Office. For this we had only one
course to follow: to work together with the consular secretary
of the American Embassy. In his book of instructions, there
were directions as to how to take care of the proper documents
and drawings, but our diplomatic friend knew nothing of engineering or of patent requirements, and Dr. Lee and I had to
interpret his own books to him.

We learned to know the embassy people very well and were
genuinely pleased by the quality of the attachés we found
there. The requirement of knowing something specific and
difficult, such as an oriental language, definitely weeds out the
playboys and the incompetents in the diplomatic service. Indeed, I think that this service has always been at its best on
its periphery.

For a long time, the frost held off, but at last the ponds and
ditches froze. Notwithstanding the short winter, there were

days quite as severe as any in Boston. It was too dry for snow, and the favorite sport of the good Peiping citizen was to go skating on Japanese-made skates on the ponds in the Forbidden City. It was quite a sight to see a sparsely-bearded elderly gentleman in a long fur-lined gown glide over the lake, doing a figure or two.

Of course, the working people, both men and women, wore the short gown and trousers, thin in summer and wadded in winter. The poor students and people of sedentary life but of humble status wore the long blue or gray cotton gown. The modern men of the university wore European costume in spring and in autumn, but in the summer they preferred the greater coolness of the thin silk gown, often worn over European trousers and shoes. In the winter, the chillness of the corridors and classrooms made the fur-lined gown almost a necessity.

My daughter Peggy had another attack of the middle-ear infection which had so often made her a victim in her childhood. There was nothing to do but to send her to the American hospital in town and leave her there with a very competent Chinese ear doctor and excellent Chinese nurses.

Meanwhile we had a chance to appraise the faithfulness of our number-one boy. It was Chinese New Year's, and of course we had given all of the servants the day off, as one would at Christmas at home. The number-one boy refused to take the day off because a member of the household was sick, even though that member was under adequate care seven miles away and he could do nothing about it. I think he liked us, but I think it was even more a sense of honor and professional duty which constrained him to this course. He was very angry at the maid for taking the day off and never quite forgave her, which caused an eruption of the quarrel, traditional in China, between butlers and maids.

Among my Chinese friends were scientific skeptics in the

modern sense. Some were Christians, but almost none were Taoists or Buddhists of any devoutness. Yet there was common to almost all that love of the whole world rather than of any specific humanity, which is so characteristic of Buddhism. Equally characteristic of China is the light, epicurean play of thought which has attached itself to the rather quaint and formless body of Taoist legends.

All the good Chinese I have known carried on the Confucian traditions, and they were not the less Confucians when they were Christians; for the Chinese have a religious tradition of syncretism, and for them the appreciation of one religion does not mean the rejection of others. Behind all Chinese who respect any religious tradition whatever is the Confucian concept of the gentleman-scholar-statesman: of a rather austere and courteous personality, tempered by a sense of humor, with the welfare of the community as his goal and dignified scholarship as his means.

There are many ways to be evil, but there are also many sources from which a good life may spring. The Confucian character is a very interesting and tempting source of the good life, and few of the more sensitive and intelligent missionaries have been able to return from China without a deep understanding and participation in the Confucian point of view. China converts its converters.

Even those Christian Chinese who are fully aware of the fineness of the better missionary and of the manifold things he has done for education, medical help, and social improvement do not welcome the missionary as an institution. Their ideas of a Christian church in China is that of a Chinese Christian church, rooted in the people, with a Chinese clergy. They resent the protected extraterritorial position which the missionary has held until recently, and they resent equally the tendency of the conventional missionary as an institution to talk down from above.

I think that Yenching was one of the finest monuments of an enlightened missionary movement, which was already departing from the purely missionary standpoint and taking on that of a native Chinese Christianity.

There was a broad and durable companionship between many of the Westerners and many of the Chinese. Yet the Chinese all felt strongly that they should be masters in their own house and that this transfer of authority should involve no avoidable delay.

Even those Westerners who loved China and wished to help it were appalled by the impossibility of raising the Chinese standard of living to a Western level without making its price beyond anything which a bankrupt country could afford. Studies made at the Peiping Union Medical College had shown that the pitifully inadequate diet of the Chinese peasant, which was shown in the lean stomachs and the spiderlike limbs of the rickshaw boys and the peasants, could not be improved into something more nourishing at any price which the peasants could pay.

Corruption and graft, which seemed an irremovable part of Chinese life, were not to be put aside by any arrangement which appeared practicable during the time of transition. If you destroyed the customs by which the Chinese favored his family and his clan against the demands of political and business honesty, millions of people would have been doomed to starve before any new order could come to replace the old. It is quite understandable why the Chinese should look longingly toward any shortcut to modernization, industrialization, and a higher standard of living which might be offered them.

We went on immersing ourselves deeper and deeper in the life in China, and sometimes were a little eager for variation in the small circle of our companions. My friend Professor Hadamard, of Paris, came at the beginning of the second term. He was installed near us in the Old South Compound, but he soon moved to an apartment in the city in or near the legation quarter. The Hadamards were much happier in this livelier environment. Professor Hadamard was already well along in years, and the discomforts of the isolated life on the grounds of the university rather terrified him. It was a pleasure to see him again. He was a great mine of reminiscence of the good old days of French mathematics. His wife was also a mine of anecdotes concerning French academic life, and she had known Pasteur when she was a child.

Hadamard told us a delightful story about his own youth, when he felt a certain fear of disfavor from his more conservative colleagues because of his kinship with the wife of Colonel Dreyfus. With the Dreyfus affair exciting all France to great emotional heights, everyone was either an ardent Dreyfusard or an ardent anti-Dreyfusard. Among the latter was the great mathematician Hermite, who was to examine the young Hadamard for his doctorate. Hadamard approached this occasion with fear and trembling, and this embarrassment was not relieved when the old gentleman said to him, "M. Hadamard, you are a traitor!" Hadamard mumbled something in confusion, and Hermite went on to say, "You have deserted geometry for analysis."

We used to go to town to visit the Hadamards and sometimes Margaret and I, or Lee and the two of us, used to go down into the tangled, squalid streets of the so-called Chinese city (as opposed to the rectangular Tatar city) to rummage in the antique shops. There we would often come across ancestor portraits which show dignified Chinese gentlemen or ladies, in stiff poses, with hands on the knees, dressed in marvelous silken gowns, which for the men were robes of office, civil or military. For all their pomp and stiffness, it was common for the faces in these pictures to be of a remarkable fineness, humor, and sensitivity.

We found one such ancestor portrait which was so like Professor Hadamard himself, with his somewhat sparse, stringy beard, his hooked nose, and his fine, sensitive features, that it would have been completely adequate to identify him and to pick him out of a large assembly of people. There was, it is true, a very slight slant to the eyes, and a very slight sallowness of complexion, but not enough to confuse the identification. We bought this picture and gave it to its likeness. He appreciated it very much, but I don't think that Mme. Hadamard cared for it. She did not wish to think of her husband as a magot: the conventional nodding figure of a mandarin which one finds in French tea shops, and which occupies a conspicuous position as the emblem of the famous Café des Deux Magots.

At any rate, in their later wanderings the Hadamards managed to lose or to misplace this portrait, and thus my more recent visits to the Hadamards in New York and in Paris have not enabled me to verify again from direct observation the resemblance between Hadamard and the Chinese picture.

As I have said earlier, I had long wished to meet Mandelbrojt face to face and to discuss with him the relation of our researches. Hadamard told me that he too was eager to talk these matters over with me. This was all the more important as he had taken an active part in the organization of the International Mathematical Congress at Oslo, which I was to attend that summer. Hadamard wrote to Mandelbrojt and made arrangements for me to visit him on my trip to Oslo and possibly put in a few days of joint work with him.

This habit of joint work is almost the peculiar property of mathematicians and mathematical physicists. Most other scientists are hampered by the fact that they are dependent not only on a laboratory but on a very special laboratory with their own materials and equipment. Historical and philological workers usually find themselves in so controversial a field that a joint paper would scarcely be possible, unless by an extraordinary accident they should happen to share not merely the same general views but even the same detailed opinions. In the arts such as literature and music there is not enough common ground to enable a group of artists to achieve together that unity of personal point of view which is essential for true creative work. Mathematics, however, for all the real, personal individuality of point of view which characterizes the aesthetic side of that discipline, is sufficiently factual for collaboration to be possible and for differences of opinion to be judged and dissolved on a non-personal basis.

Eventually the time began to draw near for us to leave China, and we began to consider how we should do so. My main motive for going home by way of Europe was that I wished to participate in the Congress in Oslo. We had thought of going by the Trans-Siberian Railroad. I took council with the officials of the Soviet Embassy to see if it would be possible. They said it was, but Margaret and I eventually came to the conclusion that the two weeks' trip on a train would be too hard on the children and we decided to take the longer sea trip by the Indian Ocean. We found a Japanese boat which was nominally first class but actually second and was not too expensive. We made plans to embark in Shanghai.

We took the train from Peiping, accompanied by Dean Ku, of the Tsing-Hua Engineering School, and a Chinese friend of his. The fact that the cars were divided up in European fashion made it much easier to take care of the children. The dining car offered us a choice of two styles of meals, Western and Chinese, and we unhesitatingly chose to eat Chinese. The next evening we arrived at the train ferry on the north bank of the river opposite Nanking.

Ku and I left my family on the train and took the swifter passenger ferry to the city, where we drove over to the house of my old friend Y. R. Chao and found not only a family party but a number of interesting officials from the university and some officials of the government. The two American-born Chao girls were now making their appearance as individuals, and there were two younger daughters, who had been born in China.

In a few days our boat was in—the *Haruna Maru* of the Nippon Yusen Kaisha. The captain was a very amiable gentleman who spoke excellent English and who, I believe, had spent some time in America. He was very solicitous for our interest. Indeed, although the ship was small and crowded and not particularly well equipped for the tropical service in which it found itself, the spirit on board was very pleasant and we rather enjoyed the trip. We found ourselves immersed for six weeks in a world of which we had heard only through the writings of Somerset Maugham.

One new passenger of great interest came aboard at Hong Kong. It was Father Renou, a French Lazarist missionary priest from the interior of Yunnan, who was on his way back to the headquarters of his order in France to report at first hand on the needs of the China service. We found him a charming and intelligent companion. He felt that there had been sent out too many priests who were French peasant boys full of good will and sanctity but ill-trained to compete with even the fragmentary scientific learning of the Chinese village school teacher or to fulfill the magistrative duties which the treaties had conferred upon them.

Politically, Renou was a liberal who had very little sympathy with the Fascist reaction which had become so strong in Italy and had met with a certain sympathy among a sector of the clergy. He was a historian, and was greatly interested in the way the Church had lost a magnificent opportunity to embrace the whole of China toward the end of the seventeenth century. This was, he told me, the Affair of the Rites, when a great many Chinese would have agreed to embrace Catholicism if only a Christian interpretation could have been found for their traditional ceremonies.

Professor Fujiwara, of Sendai, was on the boat, bound as we were for the International Congress at Oslo. There was also a Chinese lady on her way to marry her fiancé, at the embassy at London. The Chinese lady was very shy and almost a myth to us on board, but Father Renou tried to help her in her loneliness by having Margaret come down and talk Chinese with her. Apparently the lady had received instructions not to confide in anybody until the boat landed, and all my wife's overtures were in vain. However, after I had left the boat at Marseille and they had passed Gibralter, the lady had to come on deck to fill out the documents necessary for landing. There she attached herself to Margaret and remained close by her side until the boat landed and her betrothed appeared to take her off.

We played a great deal of bridge on board, and I showed myself to be a habitual overbidder, thus incurring the wrath of some of the crustier experts. I also played some chess. Here I initiated into European chess one of the Japanese officers who was manifestly an expert at the very different chess of his own country, in which the pieces are wedge-shaped and their direction rather than their color indicates the side to which they belong. It is worth noticing that in Japanese chess, captured pieces may be used by the captor—a fact which may not be without importance in considering the Korean War. How else can we interpret the use of the Korean prisoners of war as pieces still in action against us, or as our prisoners of war as pieces to be turned to the disadvantage of our enemies? At any rate, my Japanese friend learned Western chess in two or three sittings and proceeded to demolish me with an absolute consistency.

Beside these amusements, I made a first attempt at writing

a novel, based on some of my own experiences in the academic world and on some characters I had known. The novel was very amateurish, but it helped fill in the interstices of boredom in the long trip. In addition, I got the practice of at least trying to write on fictional and human material, and this practice has stood me in good stead ever since.

While the boat was tied up in the Suez Canal, Margaret and I made a brief trip to Cairo. In a few hours after our return to Port Said, we were drinking in the cool breezes of the Mediterranean. The next day we passed through the Strait of Messina, between Sicily and the mainland of Italy, the home of the magic Fata Morgana.

We emerged into the Tyrrhenian Sea, and soon the active volcano of Stromboli loomed before us. This was the captain's first trip to the Mediterranean and he was as interested as any of us at seeing the sights, so he had the kindness to swing the ship all around the mountainous island before continuing on his course to Marseille.

I debarked at Marseille alone, leaving the family to continue the boat trip by way of Gibraltar and the Bay of Biscay to London. I took the night train to Clermont-Ferrand. Mandelbrojt met me at five in the morning.

He was exceedingly friendly and cordial, even though this was our first meeting. He was a Polish Jew who had come to study and live in Paris and had attracted the attention of dear old Hadamard. Hadamard's accolade was enough to make a young mathematician, and Mandelbrojt was well launched on a fine career. We put in four days of hard work on a joint paper which we were later to present at the Oslo congress.

There was a general strike in France at the time, and Mandelbrojt was, on the whole, sympathetic with the strikers. Nothing happened to molest us. I lived at a small hotel but passed the day in Mandelbrojt's house, working with him and dining with his family. Mandelbrojt took me around to see the sights. Finally I left for England a few days ahead of the arrival of my family.

In London I looked up the Haldanes, who were now living in an interesting quarter of town near Regent's Park and the Zoological Gardens, in which Haldane took a great interest. Mrs. Haldane welcomed me cordially, but I could not see the family that evening as they were giving a party for H. G. Wells. However, the next day I visited them, and Haldane showed me some of the particularly interesting specimens at the zoo which were not generally shown to the public.

In a couple of days I went down to the docks to pick up my family. We began at once to plan how we should settle the children for the duration of our visit to Oslo. We found a sort of vacation school at Bexhill, on the south coast, where we left them in good hands for a few weeks.

Margaret and I took the rough and fatiguing North Sea trip to Denmark and spent a day or two in Copenhagen visiting friends. The city had become much more a place of rush and of automobiles than we had found it on our last visit, and its quaint and homelike flavor, though it still remained, was now considerably diluted. We took the train for Oslo, crossing the sound by train ferry from beside the jutting castle of Elsinore to the Swedish city of Hälsingborg. We found many American and European mathematical friends on the train.

The southern part of Sweden, through which the train ran, made me think very much of the Maine coast. Even the houses were wooden. The rocks themselves had the familiar rounded, glaciated appearance we so well knew at home, and we felt as familiar with the country as if we had been there before.

Our Danish stood us in good stead and enabled us to converse even with our Swedish fellow passengers. When we crossed into Norway the language difference was still smaller, and at no period did we feel the helplessness of the tongue-tied monoglot.

The meeting was delightfully arranged, and as exciting and overstimulating as such meetings always are. We found ourselves among a host of friends of all nationalities. We joined up particularly with our colleagues the Vallartas (for Vallarta had married since our early days, and Mrs. Vallarta became a good friend of my wife). Canon Lemaître, of Louvain, also belonged to our group. He had worked for years at the Harvard Observatory. We found him in Oslo, as we had known him before, a delightful friend and table companion. We would go with him on excursions and walks during the twilit midnight of an Oslo summer.

Margaret left after the congress to see her kinsfolk in Germany. I went back to England, for I had planned to spend a couple of weeks with the Haldanes in the charming county of Wiltshire. The Haldanes had found a delightful old stone house among the rounded hills and steep valleys of that region. Haldane had ascertained that his wife loved the downs of Sussex, but the downs of Sussex had become too expensive for a university professor. He took the geographical map of England and followed the chalk formation of these downs around until he came to a somewhat less popular region where he could count on a similar landscape. The rest was easy, and the Haldanes had settled in a country of delightful walks and views.

Both Haldane and myself were grieved and depressed at the new blow to freedom in Spain. Later on, Haldane offered his services to the Spanish republic and left to fight there. In Spain he was appalled by the bumbling inefficiency which accompanied the good will of most of the anti-Franco parties, and more and more he began to lean toward the Communists. According to him, they at least had a purpose and a policy.

The English Communists soon recognized in him one of their greatest assets. He continued to have an editorial post on the *Daily Worker* until his break with Communism on the basis of the dogmatic biology of Lysenko and the Czechoslovak trials.

However, this political rapprochement with Communism was yet to be. For the most part our talks on our long tramps about the bare-topped hills were scientific or literary. I showed him the fragment of the novel that I had already begun to write, and he showed me the manuscript of a series of children's stories which he was soon to publish under the title *My Friend Mr. Leakey.*

Looking back on my China trip and on the European visits which ensued, I see now how much progress I had made since my earlier years at M.I.T. I had a successful family life even if I was not the easiest of husbands and fathers, and Margaret and I now had a large stock of common experiences to enjoy together. My children were more than babies and had begun to be companions. They were starting their lives with the enormous moral advantage of seeing the world as a whole and not merely as an interplay of master races and servant races. My scientific career had reached the point where my accomplishments were unquestioned even if they were not yet welcomed in certain quarters near home. I had begun to see the fruition of my work not merely in a number of important independent papers but as a standpoint and as a body of learning which could no longer be ignored. If I were to take any specific boundary point in my career as a journeyman in science and as in some degree an independent master of the craft, I should pick out 1935, the year of my China trip, as that point.

11

The Days
before the War.
1936—1939

During my last few days in England, while I was visiting the Haldanes in the West Country, Margaret was still over in Germany. When she returned, we picked up the children at the south-coast summer resort where they were staying and left for home. One year's absence had made it necessary for us to take up anew many old threads in our M.I.T. life.

I came back to find the situation in the mathematics department rather confused. Besides Eberhard Hopf, who had been a member of the mathematics department for several years, we had another young man, Jesse Douglas, working with us. Douglas had just completed a brilliant piece of work on the form of minimal surfaces such as are generated by soap films. This is a classical problem and Douglas had so advanced it that he had won the Bôcher prize, which I had previously received for my work on Tauberian theorems.

It must be borne in mind that the depression had frustrated some of President Compton's attempts to raise M.I.T. salaries to the level of those of the greatest universities. We were thus faced by a dilemma. Either we could stay within our means and obtain mediocre mathematicians for mediocre salaries, or we could deliberately seek for those undervalued in the general market, with the hope that as economic conditions improved, and after Compton's program gathered momentum, we could ultimately bring their salaries up to a proper sum.

As a matter of fact, it was not many years before salaries began to go up again, and those of us who had confidence in the bounty of M.I.T. have not been disappointed. Still there was a period when those brilliant men who were undervalued in the general market and had been obtained by us at a discount felt exploited.

Thus, it was no surprise that our two brilliant young men felt aggrieved. So long as I was about, I could talk frankly on the matter and help build up hope for a better future. Once I was away in China, the two of them worked on one another's nerves. The scholar tends to have the sensitivity and, with that, the excitability of the artist. By the time I returned from China both Hopf and Douglas were so far set in their emotional attitudes that they were permanently lost to M.I.T.

It is Hopf's case that was the most interesting. He was a German of sufficiently correct racial origin to be acceptable even in Nazi Germany. Originally he was hostile to Hitler, or at least sympathetic to those on whom Hitler had wreaked his ill will. However, there were strong family influences pulling him toward the Nazi side.

When my cousin, Leon Lichtenstein, had died as an indirect result of the coming of Hitler, the Germans looked around for a successor. At that time good mathematicians were leaving Germany en masse, and a successor was not easy to find. Finally Hopf's name came up, and he was offered the position.

It must be borne in mind that a university position in the Germany of the good old days had a prestige both social and intellectual beyond any comparison with that of a similar position in America. A full professor at a German university was socially superior to the most successful industrialist. What the Nazis offered Hopf, if it could have been taken at its face value, was financially out of the class of anything we could hope to do in the immediate future and considerably greater in prestige than anything we ever could offer in the remote future.

I will say that Hopf consulted with a number of German refugees from Hitlerism and that they did not oppose his acceptance of the offer as vehemently as one might have expected. In the first place, while they were irreconcilably opposed to Nazism, Hopf was obviously willing to temporize with the movement, and they could not argue with him as if he had had any very strong feeling in the matter. Moreover, it was certainly better for Germany to have a man who, although not a vehement anti-Nazi, was at least not a vehement pro-Nazi. Many of the German refugees believed that Germany would either be defeated or by an intrinsic revulsion would sooner or later cast off Nazism, and all their opposition to Nazism had not affected their pride in Germany as such. Hopf would form part of an element in the new Germany which would be at least a possible basis for the re-establishment of academic sanity after the war.

The authorities at M.I.T. did not like to have a pistol presented to their heads in the form of Hopf's claims to an immediate promotion over the heads of older men. If the German offer was to be taken at its face value, and if all moral issues were to be ignored, Hopf was right to accept it. Of course, all of us hoped for an ultimate defeat of Germany, and we were more than suspicious that in that defeat the whole academic system and Hopf himself would go down in ruin. That was a matter for Hopf to decide and for no one else.

Hopf accepted the German offer. In his delight at his sudden rise, he was most condescending to his colleagues at M.I.T. To me he expressed his feelings that I was not getting my full deserts, and he wished that I could find such an advancement as he had found in Germany. I need not say that this condescension was not welcomed.

It is interesting to note that the refugee scientists who stayed in America have made great contributions to American science both for war purposes and for peace. More than half of the leading figures in nuclear science came from within the Axis. Here I need only mention Einstein, Fermi, Szilard, and von Neumann. Von Mises also came over later and made important contributions to statistical theory, as Courant and several members of his school did towards the introduction of the prevailing European techniques of applied mathematics.

My former student, Norman Levinson, was back from England where he had gone on a National Research Fellowship. I did everything I could to secure him for the department, but I found very mixed support for him in high quarters. There were those who helped me to the limit in backing up my judgment, but there were others who felt very definitely that we had enough Jews already. Among these was at least one Jewish colleague, who believed that the welcome he had received in the department might be damaged if more Jews came in, and who treated this welcome as if it were his personal property. As to myself, while I was quite willing to recognize that it might be good policy to distribute people rather widely by racial and cultural origin, I believed then as I believe now, that all such considerations are purely matters of convenience and should be subordinated whenever one has the chance to get a man outstanding on his own merits. Good men are too rare to allow a

school to pick and choose on the basis of secondary considerations.

The Harvard tercentenary occurred in 1936, and a good
many scholars gathered from all the world for the occasion.
Hardy was there from England, and I appealed to him to back
Levinson. His effort was successful, though his task was not
easy. Levinson has remained a member of the department ever
since, and continues to be a tower of strength there.

Thus this was a period in which I was subject to a great
number of separate emotional strains. The fact that Nazism
threatened to dominate the world was a continual nightmare
to every man of liberal feelings, and in particular to every
liberal scientist. I was able to take part of my internal storm
out in active measures to help the refugees, but this was not
enough to give me anything like peace of mind.

The old strains and stresses of my education as a prodigy
came back to bedevil me. For all my love for my father, those
near to me were not slow in reminding me that I was after all
nothing more than my father's son. The fact that I was a Jew
rendered my emotional situation somewhat ambivalent. In
America there was a reaction in our favor from the atrocity and
terror of the German situation, but this did not completely
compensate for the knowledge that somewhere in the world
we were being threatened with extermination, and that Nazi
anti-Semitism had provoked an echoed anti-Semitism in some
American quarters.

I had not only to suffer the direct stresses and strains of my
origin and of my early education but the subsidiary strains as
well, which originated from my entering academic life from a
rather unusual angle, without sufficient social maturity to know
just what I was and where I was going. These strains had been
much eased in the course of time by my marriage to Margaret,
but I am afraid that I had merely transferred to her the impact
of the conflicts already implicit in my own nature.

As the years went on, some of my difficulties decreased, for people will forgive in an older man what they will not tolerate in a youngster. Nevertheless, the epoch which would naturally have been the time of my emotional release was complicated by the strains of the depression, of Nazism, and of the threat of war, so that there was no period during which I could recover in peace from my earlier conflicts, and in which I could taste a few years of true serenity.

What with the Jesse Douglas problem, the Eberhard Hopf problem, and the problem of securing for Levinson his just deserts, all added to the tenseness and suspense of the prewar period, I was in a state of confusion. When I came back from China I was forty-two years old, and I had already begun to feel that I was no longer a youngster. The burden of many years of hard life had started to tell on me. Following the advice of my wife, I consulted a doctor friend who had gone from internal medicine to psychoanalysis.

Under the circumstances it is scarcely astonishing that I needed psychoanalytic help. Indeed, notwithstanding a deep skepticism concerning the intellectual organization of psychoanalysis, I would have sought this long before—if I had known the right quarter to which I could turn. I made some abortive attempts to undergo psychoanalysis during my stay in China. Even then, however, I began to learn that the more individual a man's background is, the harder it is for him to find the right practitioner.

Even as a child I had read on psychiatric topics, and I was familiar with some of the writings of Charcot and Janet. Moreover, my own experience had convinced me long before I had heard of Freud that there were dark lacunae and concealed urges in my soul that offered a deep resistance to being brought to light. My studies in philosophy had made the notion of the unconscious no novelty to me, and I was aware of the cruel and almost unspeakable impulses which this unconscious hides as well as of the nearly irresistible tendency to gloss over them and to bury them beneath a layer of rationalisation.

Thus, when I learned about Freud and his ideas, I was quite prepared to see in them a new revelation with a real measure of validity. Nevertheless, I was repelled by the internal rationalizations of the psychiatrists themselves. Their answers to all human problems and to me were too glib and too pat. Without denying in any way the therapeutic validity of much that they did, it did not seem to me that the intellectual roots of psychoanalysis had yet reached that degree of convincingness and scientific organization which carries with it full conviction. Furthermore, the maxims of the need for submission and for financial sacrifice on the part of the subject of psychoanalysis appeared to have too much in them that is professionally and financially advantageous to the psychoanalyst to seem fully objective.

Freud himself had obviously carried out on his own soul a large degree of psychoanalysis, without putting himself in that classical passive attitude which he himself later defined, and I had seen in my own person the beginnings of a psychoanalytical consciousness which was not imposed from the outside. I therefore was hardly ready to throw myself in the recommended state of full submission.

Neither was I ready to accept without question the orthodox psychoanalyst's valuation of personality, and the goals set by him as a result of a successful psychoanalysis. I never have valued contentment and even happiness as the prime objects of my life, and I began to fear that one of the aims of the conventional psychoanalyst was to remake his patient into a contented cow.

I performed the usual analytical reporting on the psychiatric couch and tried to supplement it by all that my insight could furnish concerning my own motivations and my internal set of values. I let the analyst know how deep I found the impulse to creative work and how much the satisfaction of success in this work was of an aesthetic nature. I also told him what my tastes in literature were, particularly in poetry. There are passages from Heine, especially in his *Disputation* and his *Prinzessin Sabbath,* that relate and express the religious exaltation of the Jew, which I cannot recite without tears. I told him, moreover, how the sudden shift in Heine's attitude between awareness of the degradation and baseness in daily life and the exaltation of declaring the glory of God and the dignity of the despised Jew, create in me a deep sense of awe.

All these things my psychoanalyst rejected as not coming from the true depths of my subconscious. For him they represented merely the things that I had learned at conscious levels and were of no importance in comparison with the slightest tag of a hint got from a half-remembered dream. Conscious they may have been, but their ability to move me did not come from any superficial level of my consciousness.

My analyst regarded them as a sort of contraband which had not paid the duty of a psychoanalyst's couch. He refused to consider them to any extent, and he left me with a deep feeling of having been misunderstood and misrepresented. He accused me of that cardinal sin of the psychiatric patient known as resistance. I certainly did resist, but the very fact of this resistance was a clue to much that I had experienced, and to much that was at the bottom of my spiritual make-up. Finally we parted, after a most futile half year of trying to get something from a man who, as I am convinced, did not have very much of a notion of what made me tick.

Later on, I have turned to other psychoanalysts who did not go so much by the dream book, and who made a far greater effort to get in rapport with me as a human being. These more sophisticated and sympathetic friends do not make so much of a fetish of the ritual of the couch. They do not omit to record my dreams and contradictions. However, they treat me far more as an individual than do their ritually orthodox Freudian colleagues. For them the psychoanalyst's couch is not a bed of Procrustes. They accept such differences of opinion as I have with them without labeling them at once with the damning epithet of resistance.

Naturally I was not in a position to confine my attention to my inner problems and reserve all my efforts for the baring of my soul. I still had a certain responsibility left in the matter of placing refugee scientists. However, this responsibility had become less arduous with the years, and their problems were assuming a new nature.

More countries, such as Finland and even China, were furnishing their quota to American science. The early arrivals were gradually depending less and less on their native language, which was generally German, and were taking up the American way of life as something normal. The older people were bringing up their children within the American tradition, and it was obvious that whatever might come, few of them would return to their homes in Europe on anything but a temporary basis. The younger immigrants were marrying into American families.

The final result of the great immigration of Hitler times is not yet to be seen, but it is certain that the contribution to our mathematical life of new individuals and new stock will prove to be comparable to that of the Germans in 1848, or to that of the Huguenots who migrated to England, to Holland, and to America at the time of the revocation of the Edict of Nantes.

Naturally, with so many first-rate scientists added to the American community, I collaborated with several of them in research projects. Aurel Wintner had come over to America, if I am not mistaken, before the great rush of immigrants, on the basis of a recommendation from his teacher, my cousin Leon Lichtenstein. One summer Wintner and his family took a cottage in New Hampshire some twenty miles from us. In that part of New Hampshire, to be twenty miles away is to be a next-door neighbor.

Professor Wintner is a very alert, enthusiastic scientist, quick in his motions and his thoughts, and highly original in his ideas. Mrs. Wintner is the daughter of the well-known German mathematician Hölder. These marriages of mathematicians to the daughters of their professors are so typical of the academic world, both in Europe and here, that there has come to be a saying that the genetics of mathematical ability is peculiar—it is not inherited from father to son, but from father-in-law to son-in-law.

Ultimately Wintner came to be a more or less permanent summer visitor to our part of New Hampshire. We began to undertake work together concerning a variety of topics in his own field. Some of these dealt with the extension of certain of my ideas concerning generalized harmonic analysis to the orbit problem and the perturbation problem in celestial mechanics. These represent a modern approach to old eighteenth-century problems of Laplace and Lagrange.

Another piece of joint work dealt with a modern probabilistic approach to Maxwell's kinetic theory of gases, treating a gas as a set of moving particles acting under mutual forces. I had already done some earlier research in this field tying up with the work of two physical chemists, then at Columbia and now at Chicago.

A third line of work which Wintner and I followed out together was the tightening and simplification of the proofs of the ergodic theorems of Koopman, von Neumann, and Birkhoff. These theorems, which I have mentioned already, had furnished the missing step in Willard Gibbs's work and allowed a rigorous carrying through of his idea of time averages with averages over all possible worlds. In this latter work we benefited much by discussions with the young Dutch mathematician, E. R. Van Kampen, who was our companion on various hikes through the White Mountains. Poor Van Kampen, who seemed to have a most promising career ahead of him, died a year or two later of a brain tumor.

Over all this period I had hoped that a return to lecturing in China would not be many years away. This hope was, of course, completely destroyed by the events of the next few years. In 1937, my colleague, K. S. Wildes, of the electrical engineering department was my successor in China. He returned just about the time of the battle between the Chinese and the Japanese at the Marco Polo Bridge.

Besides its world consequences, this battle had personal consequences which I felt very deeply. At the time of the Marco Polo Bridge incident, Lee and Mrs. Lee were visiting friends in Shanghai. The ensuing war between China and Japan caught them away from home, preventing them from returning to Peiping. For part of this period Lee was able to find teaching employment in Shanghai, but for much of the time the Lees were left to live on their own resources and on Lee's knowledgeability in matters of Chinese art.

To suffer this interruption of his scientific development at what should have been its best and most critical period was a serious loss indeed; and the problem of how to handle the situation, when any handling should become possible, distressed me greatly. I did everything that I could to bring Lee over to the United States, but at this time I had no positive success.

Wildes had been as much interested and attracted by his study in China as I had been a year earlier. The two of us were very busy in the ensuing years trying to influence American opinion to give increased aid to China. We went to President Compton for help in this effort, and he took a leading share in the China relief situation. Other high authorities of the Institute also participated in this.

While Communism itself never appealed to me, I have not

been able to feel that right views may not be held on many topics by members of a group of which I do not approve. When it was the fashion for Communism to condemn Nazism and stand up against race prejudice, the fact that these admirable opinions were held by Communists was made an excuse for rejecting them that only a fool could maintain. That after the defeat of Nazism the Communists have become the chief fear of the West and that they have behaved with much of the tyranny of the aggressors that they have replaced does not change in the least the fact that some of the things they stood for in the period between the wars belong to the attitudes of every decent man.

The change in Communist attitude and tactics is quite a sufficient reason for rejecting them as mentors, and certainly for distrusting their intervention; but it does not in the least alter the fact that in the period of confusion when no single party had clean hands, many young men tended to look toward Russia. For some of these, the participation in radical movements was an important stage in their moral growing up. It taught them not to take out their grievances against the world in mere sulking but to try to do something of public benefit. This habit of active participation in moral issues has long outlived the period when they believed in Communism.

Under these circumstances it is easy to see why I did not shy away from a request to help the Chinese cause, even when this cause was backed by armies not attached to the interests of the bumbling, inefficient Kuomintang.

For some time we felt confident that the help which the United States was giving to China was going through the right channels. However, disturbing rumors began to reach us. I gradually began to hear statements from informed Chinese and from traveled Americans to the effect that the Kuomintang was a broken reed, that it was not using American aid effectively, and that it was diverting large shipments of arms and of medical supplies to be resold by the more corrupt members of the party.

About this time, I was approached by a group of people who wanted me to sponsor aid for the Communist soldiers in China, who seemed to be doing a more competent job in fighting the Japanese than did the soldiers of the Kuomintang. I accepted the invitation because I felt it to be to the advantage of the United States.

The people who were really running the movement were a sincere but undistinguished and inefficient lot, and they seemed to me to be wasting more time in grubby social gatherings and in talk than in the effective collection of money. I later found out that the group contained some who acted as if they belonged to the more peripheral ramifications of the Communist Party, together with a rather larger group of well-intentioned and stable friends of China.

No alert person could have gone through the years from the depression to the beginning of the Second World War without some experience of the American repercussions of Communism. The young men coming up in academic life during the depression period were forced to recognize that they were stepchildren of the present world order. Security had become a dream of the past, and the various hatreds which made up the complex of Fascism, Nazism, and Ku-Klux-Klanism were an ever present threat, particularly to members of out-groups. They were seeking some movement or some attitude to which they could attach themselves, and they were bound to hear among other voices those of their convinced Communist colleagues.

From the very beginning I had been repelled by the totalitarianism of the Communists, even as I have always been repelled by the whole apparatus of orthodoxy and conversion in whatever religion it might occur. However, my very need for independence made me loath to interfere dogmatically with the decisions of the young people about me. In an age of many bigotries, I certainly could not feel the bigotry of the Communists as our greatest immediate threat, nor could I fail to recognize that the appeal which Communism had to some of my young friends was an appeal to their humanitarian instincts.

In the course of time, almost all of them have seen the power-politics side of the movement and the way in which good intentions have been turned into tyranny by ambitious men. They are not now Communists, nor have they been for many years. However, the effect of the fear of a witch hunt against Communism was to make it difficult for them to find an honorable and self-respecting way out and, if anything, tended to delay their exit from the Communist milieu.

My attitude toward the Chinese was reinforced by similar American support of the Spanish Loyalists. Here the moving spirit was Professor Cannon, of the Harvard physiology department. He was without any doubt the great man of American science at that period, and he had lectured in Spain a few years before.

Spain is a country which has not abounded in great scientists in modern times, but the one field in which it has done very important work has been the physiology of the nervous system in which Cannon himself was, of course, greatly concerned. It is no wonder that Cannon felt himself very much attracted to this revival of intellectual life in Spain and that he made it his duty to rally American support behind the Spanish Loyalists. In this he was in no position to reject aid from any quarter, and it is not surprising that an appreciable part of his aid came from Communist circles. This led to certain whisperings against Cannon, but he was far too sincere and forthright a man to be scared away by whisperings. I had joined in with Cannon in his support of the Loyalists, and it seemed to me that this policy was a valid pattern to follow in connection with the Chinese matter.

The times were also complex and worrisome for me in family matters. Father had retired from Harvard a little before the time of his accident. He was a disappointed man, and the bluntness with which President Lowell had accepted his retirement without the grace of a few kind words added to his disappointment. After his partial recovery from his accident, he continued to do research work at the Harvard Library, and even to walk there from Belmont, but his activity grew less year by year.

Soon after my return from China, he began to go downhill rather rapidly, and there was evidence that he had had a stroke. He needed hospital care, but this time there was much less sign of recovery than there had been on the previous occasion. He lapsed into an agitated state of depression, in which his mind was often confused. Yet he was fully aware that he was confused and that he was losing his grip on life. His depression often took the form of what seemed to me to be really a commentary on the baleful world political events of the time.

He would speak indifferently in Russian, German, Spanish, French, and English. When he spoke in languages I knew, I could observe in him no trace of grammatical confusion and no tendency to mix the words of one language with the grammar of another. Even when he could no longer recognize me as his son, the correctness and vigor of his polyglot speech was not in the least affected. Father's knowledge of language was not merely something painted on the surface of his brain but went right through its very texture.

I visited Father often, and took him out from time to time for trips in my car. However, he was on the way down, and it was scarcely even desirable that this half-life, which was all that was left to him, should be drawn out indefinitely. Finally, in the first year of the war, he died calmly and peacefully in his sleep.

For years Margaret's mother had made her home with us,

except during one or two visits to her kinsfolk in Germany. During this period, as I have said, German became to a considerable extent the language of the household. What emphasized the role of German in our life was a little incident which occurred at one of our visits to a Boston club, Friends of China.

There was a Radcliffe student there of mixed Chinese and German origin, whose father had been a coal-mine operator in Peiping and had married the daughter of his landlady during his student days in Germany. Lottie Hu, the daughter of this marriage, was studying anthropology in the graduate school at Radcliffe. The vicissitudes of the war had cut off her income. She summoned up her courage to ask Margaret whether she might not live with us and earn her room and board and a little pocket money by helping with the family. Accordingly we took Lottie into our household, where she became the friend and companion of my young daughters.

Lottie was completely trilingual, speaking Mandarin Chinese, English, and German with equal fluency. Since German had already become the second language of the home, this continued under the new arrangement, and both my daughters made a certain amount of progress in the elements of the German language.

My daughters were in junior high school. We had the usual amount of friction between parents and children; in particular, the two of them had a certain resentment of my intellectual position. Peggy said, on more occasions than one, "I'm tired of being Norbert Wiener's daughter; I want to be Peggy Wiener." I made no attempt to force the children into my frame, but the mere fact of being what I am inevitably subjected them to a sort of pressure with which my will had nothing to do.

I was proud of them, but I did not bring them up to be infant prodigies. I was particularly proud of Barbara on one occasion when she had read in her textbook some comment on the Latin Americans and she said to me, "Daddy, the author of this book seems to be very patronizing to the Latin Americans. Don't they hate it?" I replied to her, "You're damn tootin', child, and how!"

About this time there was a radio program established in Boston on the model of "Ask the Children." Barbara took part in this. I am not at all certain of my wisdom in having let her participate, but she did well, and she learned the elements of the art of handling herself before an audience. I have kept a degree of interest in the later destinies of more than one member of this children's panel, and they seemed to have done uniformly well and to have suffered no real disadvantage from the adventure.

Thus, like all families, we had our problems to consider and our decisions to make. I am neither certain of the correctness of the policies I have adopted nor ashamed of any mistakes I might have made. One has only one life to live, and there is not time enough in which to master the art of being a parent.

The bringing up of young children is not easy, but it lightened our household work to a considerable extent to have it shared among three women. My wife's mother was always doing little bits of work around our houses in Belmont and in the country, and she had accumulated a remarkable set of tools and gadgets for these tasks, which she so enjoyed.

She was a country-bred German woman, with a romantic outlook on life which had led her to seek the open spaces of the American West. Here in New England our farmhouse was as much a delight to her as to any of us.

It was in the summer of 1939, just before the outbreak of the Second World War, that she died quietly in her sleep in her upstairs room in this New Hampshire house. We buried her in a little country graveyard open to the sweep of the winds blowing across from the Ossipee mountains. We chose a tombstone of a pattern traditional in these New Hampshire burying grounds, but on it we had carved an inscription suiting both her

German origin and the massive vigor of her character. It was the beginning of Luther's hymn:

"Ein' feste Burg is unser Gott"—"A Mighty Fortress Is Our God." I am grateful that she was spared the full terror and humiliation of the Second World War, which was about to begin.

Even before the arrival of the war itself, the somber train of catastrophes had begun. The fall of 1938 was marked both by Munich and by the first of the series of West Indian hurricanes which have come to plague Boston of recent years. From that time on, all of us expected the world war to break on us. It held off until the summer of 1939.

That summer, after the death of my wife's mother, Margaret and I had taken a little trip together into Canada. This trip furnished the precedent for a similar motor trip which we make almost every summer. Later on I took another trip into Canada alone, this time to participate in the meeting of the American Mathematical Society in Madison, Wisconsin. I drove from my farm by a route north of the Greak Lakes, arriving in Sault Ste. Marie, Michigan, on the evening of the second day.

Here I learned that the war had already come. It was an experience curiously reminiscent of that time twenty-four years before, when the First World War had come to me on another trip, as a passenger on a German boat in the middle of the North Atlantic. All the fun and jollity of the summer mathematical meeting was exploded. We had hoped that this meeting would serve for the planning of an International Congress to take place in the United States during the summer of 1940, but this plan had to be put on ice for ten years.

I drove back East with an English colleague, and we had the opportunity to take stock of our emotions and expectations while picking grapes for a friend in New York State.

12

The War Years.

1940—1945

I came back to M.I.T. in the fall of 1939 and took stock of the situation. It was not hopeful. The passive alliance between Russia and Germany had dashed cold water on our wishful thinking that the Nazis were going to be held back in the east, and although the two countries finally fought each other, we did not at that time expect such luck. Moreover, the good will that had been building up for Russia, largely because we could scarcely see from what other direction a blow might come which could limit the Fascist aggression, was much dampened by Russia's aggressive policy in Finland.

In academic and technical circles, most of us were aware that a world war would ultimately engulf the United States as well as all other important countries. Therefore we began to turn our thoughts towards finding out in what sector of work we might make ourselves of use.

The possibilities of active military service had never been very real to me on account of my nearsightedness, and the passing of the years had not increased my physical fitness. I have never fancied myself as an administrator, nor, as a matter of fact, has anybody else ever fancied me in that capacity. It seemed obvious that I should have to turn to some sort of scientific research work.

I had had an apprenticeship in ballistic computation in World War I. Ballistic computation consists in the making of tables for artillery and small-arms fire which give the range of the weapon and various other related constants in terms of the angle of elevation of the gun, the powder charge, the weight of the missile, and so on. This work had trained me in much more than purely ballistic matters, for it had left me with a pretty general sophistication in the ways of the computing room. I had also spent much time in recent years working with electrical engineers. Therefore I foresaw that my destined berth in the war would be some sort of job in which I should apply computational techniques to electrical engineering problems. Furthermore, my work with Lee had given me a look-in on problems of engineering design.

All this was clear, but what was not clear was the direction from which the call would come. As the nerve-wracking wait of the "Sitzkrieg" began to give way to a more active and threatening military schedule, most of us came to see that the main problem before America would be to keep England in the war as an effective combatant until such time as we might enter it ourselves. This meant that the submarine and the bomber campaigns were the two chief menaces which we should help to conquer.

Luckily, England herself had given us the best possible lead for helping her in these fields, in the brilliant invention of radar. M.I.T. was pushing this sort of research from the very beginning, even before the start of the European war and long before we got into it. But for the time being this seemed to be a matter for specialists, and I was not a radar specialist.

The stream of refugees from Germany speeded up for the moment and then ceased altogether. These last driblets of immigration did not seem to me to consist altogether of persons of the same moral value as some of those who had come before. More than one of these last drippings of the wine press showed an eagerness to indoctrinate us in the irresistible momentum of the Nazi advance. Their zeal could scarcely have been exceeded if they had been paid propagandists. At last it became quite clear to us that in addition to the great cultural crop of fine, persecuted men and women who have enriched our intellectual life, there were those whose main objection to Nazism was that they were excluded from it. However, the new summer vacation came along in its good time, and we tried to make our life as cheerful as possible, notwithstanding the catastrophe about us. One cannot live in a perpetual atmosphere of gloom.

The Inghams, from the English Cambridge, were caught in America by the war and became our summer neighbors. They shared with us the pleasures of mountain walks and bathing in Bear Camp pond.

We had a curiously interesting visit that summer from the Hungarian mathematician Erdös, the Japanese mathematician Kakutani, and the English mathematician Stone. They had just got into trouble on Long Island by inadvertantly approaching too close to a radio-beacon station. They were put in jail overnight as suspicious aliens but were released later when the authorities were able to reach their sponsor, Professor Veblen, of Princeton. It was just after this contretemps that they drove up to New Hampshire, and we had a most pleasant little session on our porch. At present, Kakutani is teaching in the United States, but Stone and Erdös have gone back to Europe.

At the end of the summer, Ingham returned to England as he had planned, but his wife and children and their maid stayed again in our valley the next year. Again we went for long walks together, the children now participating in them more easily. I have seen the family several times since their return to England, where I believe one boy is going to the university and the other is an officer in the Air Force. They still retain a real love for New Hampshire and our valley.

Wintner remained our summer neighbor. He and I had planned to work together during the year 1940–1941, and he came to Cambridge to do so. It was unfortunate that my attention was concentrated on war work at the time. I feel that I was to a certain extent unjust to Wintner in not living up to our informal contract, for he found himself able to ignore the pressures of the warlike atmosphere. I could not, and though I was willing to work with him with a part of my attention, I was unable to devote to it my full interest. Thus we went our two ways, which gradually drifted farther and farther apart.

By spring, the catastrophe of Norway had occurred and the catastrophe of France was threatening. The emotional solace of our country house, in which we had become used to take refuge from the buffets of the outer world, was of no avail to us when we were confronted by the imminent loss of European civilization.

In August 1940, the summer meeting of the American Mathematical Society was at Dartmouth. It was as pleasant as a meeting could be when nothing but the war could really command our attention.

The algebra of complex quantities is vital for telephone engineering, and the Bell Telephone Company instrument, a numerical computer, was made to fill a definite need in such work. Its importance derives from the fact that our Arabic notation for numbers gives to the number 10 an artificial position justified by custom alone and constituting no part of the real foundations of arithmetic. Instead of writing a number as so many units, so many tens, so many hundreds, and so on, we can just as easily write an integer as a sum of ones, twos, fours, eights, and so on. In this case, instead of the ten digits of our conventional scale of numeration, we need only the two digits, zero and one.

The Russian peasants use what amounts to a scale of this sort, known as the binary scale, for their addition, multiplication, subtraction, and division. It has the great advantage over the scale of ten that the multiplication table reduces to the statement that one times one is one.

For obvious reasons, it is easier to mechanize arithmetic on the binary scale than on the scale of ten, and the Bell System instrument accordingly employed the binary notation. The sole serious disadvantage in doing all arithmetic on the binary scale is simply that we have adopted the decimal system and the bulk of existing numerical results are given in this tradition. When we have a large amount of new computation to do, it is often worth-while to ignore this fact and to translate all our initial data to the binary scale and all our final results back to the decimal scale.

One place where the binary system of numeration is used is in the employment of gauges for the measurement of the thickness of a mechanical part. Suppose that we have one accurate gauge of a thickness of one inch; one accurate gauge of a thickness of two inches; one accurate gauge of a thickness of four inches; and one of a thickness of eight inches. Then we can combine these to give accurate measurements by inches from one inch to fifteen inches. The code is the following: we combine our gauges on top of one another in these combinations:

1 inch	1 inch gauge
2 in.	2 in. gauge

3 in.   2 in. gauge and 1 in. gauge
4 in.   4 in. gauge
5 in.   4 in. gauge, and 1 in. gauge
6 in.   4 in. gauge and 2 in. gauge
7 in.   4 in. gauge, 2 in. gauge, and 1 in. gauge
8 in.   8 in. gauge
9 in.   8 in. gauge and 1 in. gauge
10 in.   8 in. gauge and 2 in. gauge
11 in.   8 in. gauge, 2 in. gauge, and 1 in. gauge
12 in.   8 in. gauge and 4 in. gauge
13 in.   8 in. gauge, 4 in. gauge, and 1 in. gauge
14 in.   8 in. gauge, 4 in. gauge, and 2 in. gauge
15 in.   8 in. gauge, 4 in. gauge, 2 in. gauge, and 1 in. gauge

This is equivalent to writing the numbers 1 to 15 in the following way: 1, 10, 11, 100, 101, 110, 111, 1000, 1001, 1010, 1011, 1100, 1101, and 1111.

I do not remember whether it was before or after the Dartmouth meeting that Vannevar Bush had sent around to the various members of the M.I.T. faculty a questionnaire asking for suggestions concerning the mobilization and use of scientists in case of war. This was a matter on which I had very definite opinions, and I was strongly in favor of a scientific collaboration which would cross the frontiers between one science and another, and which should at the same time be voluntary, thus preserving a large measure of the scientists' initiative and individual responsibility. I distrusted all plans that might depend on a high degree of subordination of individuals to a completely authoritative setup from above, which would assign each man the narrow frame within which he was to work. I suggested therefore the organization of small mobile teams of scientists from different fields, which would make joint attacks on their problems. When they had accomplished something, I planned that they should pass their work over to a development group and go on in a body to the next problem on the basis of the scientific experience and the experience in collaboration which they had already acquired.

But nothing resulted. Those who work almost exclusively with gadgets tend to develop a love for them, since they lack the unpredictable factors which affect the operation of the human being.

Gadgeteering very easily becomes a sort of religion. Luckily, the vicissitudes of the last twenty years of gadgeteering have somewhat shaken the faith of many men, including Bush, in the unlimited scope of the machine. However, there remain many people who have not been as directly confronted as Bush with the disadvantages as well as the advantages of machines, and these people follow the tendency of the day to favor the big laboratory and the big administrator.

On the way back from the meeting, I began to discuss with Levinson, who was now a full-fledged colleague, the general problem of computing machines, and to wonder whether this might not be the field in which I was destined to do my war work. For some time I had been considering for Bush the use of machines to solve systems of partial differential equations and I felt that a sort of television scanning would be the proper basis for the mechanization of partial differential equation problems. My new experience with the binary machine convinced me that electronic binary machines would be precisely the devices required for the high speed of computation needed in partial-differential-equation problems.

I saw that for a machine to work properly on partial differential equations, it would have to go through an almost incredible bulk of work in an almost incredibly short time. This suggested to me that the future of high-speed computing machines for these particular purposes could not lie in Bush's models, which represented physical quantities by electrical or mechanical quantities, but rather in some enormous extension of the ordinary desk computer, working as I have said, on a scale of two rather than on a scale of ten.

Now that I had come to take a serious interest in problems of the speed of computation, I was forced to consider the relative merits of two grand strategies in computational methods. One of these, which Bush followed, was called analogy computation, in which the numerical digits of computation are represented as measurable physical quantities. The other, the digital mode of computation—which is that of the desk computing machine—represents a number by the sequence of its digits.

The important point in the distinction between the analogy computers and digital computers is that the latter do essentially what we ourselves do with arithmetical operations on paper. When we represent a number as 56, we mean that it is a combination of five tens and six units. When we multiply this by 38, which is three tens and eight units, we go through the operations indicated in the following table:

$$
\begin{array}{r}
56 \\
38 \\
\hline
48 \\
40 \\
18 \\
15 \\
\hline
2128
\end{array}
$$

We never have to go beyond the multiplication table and the simple rules of addition, nor do we represent our 56 and our 38 as quantities such as 56 degrees or 38 inches.

There are digital multipling machines in which ten plays no role, which operate on the binary scale. This is how they work: Let me consider the operation in which $7 \times 5 = 35$.

$$7 = 4 + 2 + 1$$
$$5 = 4 + 1$$

In the scale of 2, these statements are equivalent to writing 7 in the form 111, and 5 in the form 101—meaning that 7 is one 4, one 2, and one 1; while 5 is one 4, no 2's, and one 1. When we multiply these we get the following schedule of operations:

```
  111
  101
 ─────
  111
 111
─────
11211
```

If we remember that, in our scale of notations, 2 = 10, the number 11211 may be written 12011 or 20011 or 100011. The last representation is the truly binary one, which uses no digits other than 0 and 1. Here this means $32 + 0(16) + 0(8) + 0(4) + 2 + 1 = 35$. This method is called the method of multiplying on the scale of two. I repeat, it is quite as digital as ordinary multiplication on the scale of ten.

In contrast, a particular analogy-computation instrument proceeds as follows: In an electrodynamometer, two coils are attracted to one another in proportion to the product of the current carried by these two coils, and this attraction can be measured by an appropriate sort of scale instrument. If, then, one coil is carrying seven units of current and the other five units, the scale will read something proportional to thirty-five. This sort of instrument for multiplying is known as an analogy instrument, because we are replacing our original situation, in which certain quantities are to be multiplied, by a new situation, in which we set up two currents in analogy to the original quantities and read off the product by a physical situation analogous to the original one.

Digital computing machines thus differ from analogical computing machines in that they can theoretically be read to the complete degree of precision of the numbers introduced, while analogical machines are restricted to the degree of accuracy with which the original situation is truly analogous to the corresponding situation, which we use to replace it in our computation. The Bush machines for solving differential equations are strictly analogy machines.

As to the relative merits of these machines, the great flexibility of electrical and other measuring apparatus makes the construction of a fairly good analogy machine easier than that of a fairly good digital machine. However, when it comes to high speed or high accuracy, the advantage is all with the digital machine. There are very few physical measurements which can be made with an accuracy greater than one part in ten thousand, and this corresponds to a determination of four decimal digits and not quite fourteen binary digits.

Moreover, to take a measurement with this degree of accuracy can scarcely be a truly instantaneous process. Analogy machines are intrinsically slower than is needed to fulfill the demands of the very fastest and bulkiest computation, so that I felt that they had already reached their apogee.

When it came to the matter of digital machines, I was forced to consider the real essence of the action of such a machine. The ordinary desk computer determines the position of certain wheels on the basis of the position of certain other wheels. Each such position is a choice between ten alternatives. It is not difficult to represent these ten alternatives by ten projections on a metal wheel, but the use of metal wheels involves very disagreeable and restrictive problems of inertia and friction.

It seemed to be preferable in every way to replace the mechanical choice characteristic of existing digital machines by an electronic choice of digits. The two advantages to be expected here were the vastly reduced inertia of a stream of electrons as compared with a sequence of mechanical parts, and the greater technical ease of canceling quasi-frictional losses—that is, resistance losses—by amplification. As a result of both of these, I was quite certain that the coming high-speed computing machines would be electronic digital machines. I may say that the idea had begun to come up in various places in the literature and that, in accepting this approach to computing machines, I was simply representing the spirit of the times.

As I have said, a decimal digital machine uses a choice among ten possibilities as its fundamental decision while a binary machine uses a choice between two alternatives. I suppose that the general use of the scale of ten came from the ten digits of our two hands. Certain races such as the Mayas have apparently counted on fingers and toes together, and use the scale of twenty. It is an interesting reflection that if the human race had been constructed after the pattern of the Walt Disney cartoons, with four digits on each hand, we should probably have adopted the scale of eight, which is merely a slight variation of the scale of two, since $2 \times 2 \times 2$ is 8.

However, there is a fortuitous advantage that, while it does not recommend the scale of ten, at least makes it easier to use it than, let us say, the scale of thirteen. Computing machines of the decimal type depend on the use of wheels with ten equally spaced teeth. To construct them involves laying out a decagon—a polygon with ten angles. This is a simple problem in plane geometry, while the construction of a regular figure with thirteen sides is not.

In the case of an electronic circuit, however, the parts which are equivalent to wheels are not dependent on ordinary plane geometry, nor is ten a particularly easy number to represent. The natural choices to be made in an electric circuit will be between pairs of alternatives.

There are well-known circuits already existing with two alternative positions of equilibrium, and these are known as flip-flop circuits. About the only easy way of constructing a circuit with ten choices seemed to be by a combination of flip-flop circuits. The logic of flip-flop circuits would lead to a combination of choices among a number of alternatives—a combination which was a power of two. Thus, it seems that the only natural way to construct a set of ten alternatives is to construct sixteen and throw away six.

In the design of a machine, you pay in effort and in cost not only for everything that the machine does but also for everything that the machine might be made to do; and to use sixteen alternatives for the purpose of selecting ten represents a wastefulness of 37½%. I came to the conclusion that the high-speed computing machine for the solution of partial differential equations would be a digital electronic machine on the scale of two.

For work on a scale of two, we need to use machines which have only two possible alternatives, such as the presence or absence of a punched hole in a card. This device was already familiar in the Hollerith machines made by the International Business Machines Corporation. This particular way of writing a number in the scale of two is, however, unsuitable for a thoroughly high-speed machine. Punching a hole in a piece of cardboard is a slow process when one considers speed on the scale of millionths of a second per operation. Some such scale is needed before we can consider a computation as really high-speed, and, moreover, the problem of disposing of the used cards and of keeping a sufficiently large stock of new cards soon becomes almost astronomical.

The speed of punching the paper might be greatly increased by using an electric spark rather than a mechanical punch, but this would leave the problem of the bulk of material just as bad as it ever was. Thus, I was naturally driven to the idea of steel tape on which a magnetic mark would be made by an electromagnet. One can read such a mark at high speed and erase it at high speed, leaving the tape blank and ready for another use.

One of the main problems with such a tape is to make the marks so small that as many as possible can be kept clearly distinguishable on a given area. This requires extremely small pole pieces for the marking and reading magnets. It seemed to me that the smallness of these pole pieces might be vitiated by the spread of the magnetic field within the tape, unless the tape itself or at least its effective magnetic layer were extremely thin.

I therefore had the idea, partly on my own and partly through conversation with those of my colleagues who were better acquainted with technical developments in this subject than I was, that the best thing to do would be to treat separately the two requirements of the tape: to carry magnetization and to hold together. We might do this by imposing a thin magnetic layer on a non-magnetic material which would have all the strength. I had thought of a thin iron layer superimposed on a tape of brass or non-magnetic metal, but I had also thought—I believe through the suggestion of a colleague— of the device which now dominates the field: of a paper tape carrying a thin layer of magnetic oxide of iron.

I have recently talked with a friend at the IBM company concerning present practices in high-speed computing machines, and in particular in those which work according to what is now known as the Monte Carlo method, solving partial differential equations by an extremely often repeated process of averaging. Apparently the devices I suggested in 1940 are substantially those which are now employed.

The odds in a gambling house are extremely regular and predictable, and the Monte Carlo method consists in setting up a mathematical problem as an ideal game, in playing it out a large number of times, and in determining the theoretical winnings. The computational device which I suggested in 1940 had the same non-static character as the Monte Carlo method of the present day and also depended on the playing out of an ideal game.

I made a report of my suggestion to Vannevar Bush, but I did not get a very favorable reception. Bush recognized that there were possibilities in my idea, but he considered them too far in the future to have any relevance to World War II. He encouraged me to think of these ideas after the war, and meanwhile to devote my attention to things of more immediate practical use.

Later on I found that he had no very high opinion of the apparatus I had suggested, especially because I was not an engineer and had never put any two parts of it together. His estimate of any work which did not reach the level of actual construction was extremely low. The only satisfaction I can now get is that I was right something like ten years before the techniques to prove my ideas were developed.

Having discarded this as my task in the threatening war, I began to look around for other places where I might be more useful. At one time I had ideas concerning a mathematical and mechanical method for encoding and decoding messages. My ideas would certainly have worked, but in this field it is not enough for an idea to work. In fact, for it to have any merit at all, it must work considerably better than existing devices, or any devices easily invented.

The question of what it means for one decoding-coding device to work better than another is not simple. It can be taken for granted that any ciphered text which is sufficiently long can be decoded by a possible enemy if you give him enough time, and it also must be considered that the problem of decoding a cipher message is not necessarily trivial even when the cipher is known. A good cipher must combine a certain ease of decoding by machine, or by a recipient in the know, with a large measure of difficulty for an enemy to decode it without the help of a knowledge of the cipher.

As is usually the case when there are two different requirements for a system or an apparatus, this does not lead one to
a single apparatus but to a number that is large according
to the weight one puts on each of the two requirements. Thus,
there are easy ciphers which will be useful in the field for
messages which need only an hour's secrecy, and there will be
difficult ciphers for messages whose secrecy must be maintained for months. There will be a whole spectrum of ciphers
in between. For this reason, the design of ciphers is not a
matter into which one can go cold, without knowledge of the
existing tradition and of the practical demands for each particular use. Again I had to look around for another possible
field of usefulness. I found this in the design of fire-control
apparatus for anti-aircraft guns.

When I was a boy, fire control was conceived primarily from
the point of view of coastal batteries and of battleships—that
is, of gun platforms whose motion relative to the target was so
slow that there was a chance for a considerable amount of
computation by very crude manual devices before the target
should have passed out of effective range.

Even in the First World War, the airplane had changed all
this. The problem of shooting down an airplane is not, of course,
like that of lobbing a mortar shell into a fortress but like that
of shooting ducks on the wing. The duck will not stay still
while you are shooting; and if you aim your gun where you
see the duck, the bird will be considerably ahead of that
point by the time the shot arrives. You must shoot ahead of
your target, and you must estimate that amount you shoot
ahead quickly and accurately. If your estimate is not correct
you will probably not have another chance at that bird.

The result is that from the very beginning it was necessary
to build into the control system of the anti-aircraft gun some
mechanical equivalent of a range table which would automatically allow the gun the necessary lead over the plane to

make the shell and the plane come to the same place at the same time. To some extent this a purely geometrical problem, but in its finer developments it involves an improvement of our estimate of the future position of the plane itself. This must be estimated from the past positions, or at any rate from the observed past positions. The problem of predicting the future position of the plane is what the mathematicians call a problem of extrapolation.

My previous electrical engineering work had made me familiar with the theory of *operators*. An operator in this sense represents a device which will change a certain electric input into a corresponding electric output. Mathematically speaking, the operator may be represented by a transformation formula, but not all such transformation formulas lead to operators which are physically realizable. The main condition for physical reality which we must impose on an operator is that the output should involve only the past and present of the input. It will be seen that the problem of shooting ahead of an airplane demands that the realizable operator approximate the future position of the plane which could, in fact, be ascertained only by a non-realizable operator. Only a prophet with the knowledge of the mind of the aviator could predict the future position of an airplane with absolute certainty, but there are often enough, in fact, means which will allow one to accomplish the minor task of a quite correct prediction.

The mathematical processes which suggested themselves to me in the first instance for prediction were, in fact, impossible of execution, for they assumed an already existing knowledge of the future. However, I was able to show there was a certain sense in which these processes might be approximated by processes free from this objection.

I do not wish to lose myself here in technicalities which will be understood only by scientists or engineers. However, I did in fact consider certain possibilities of approximating to non-realizable operators by realizable operators. I suggested these notions to Professor Caldwell, who was normally in charge of M.I.T. work on Bush computing machines, and who was now engaged in applying these machines to war problems. After the custom of those times, Caldwell immediately put a classification on my ideas, so that thereafter I could no longer speak freely of them to anyone with whom I wished to talk.

For a trial setup for my problem, Caldwell and I were tempted to make use of Bush's differential analyzer because of the ease of assembling its parts for the simulation of a large range of different problems. In this the differential analyzer resembled a Meccano set; and in fact, when the English tried to follow in Bush's footsteps and construct a differential analyzer, they used ordinary Meccano parts with very creditable success.

We made several experimental runs with different settings of our apparatus, and we found that those which we had considered in advance to be better actually were. Our instrument was an assembly of a number of adding devices, multiplying devices, and integrating discs.

At this stage, prediction theory was made a government project, and a young engineer, Julian Bigelow, who had worked some time with International Business Machines, was assigned to the problem. This was the beginning of a long collaboration between us. Bigelow is a quiet, thorough New Englander, whose only scientific vice is an excess of scientific virtue. He is a perfectionist, and no work that he has ever done is complete enough in his eyes to satisfy him.

He used to be an enthusiastic aviator, but this sport became impossible in war time and is, at any rate, too expensive for the average man. Most accidents of private flying are not serious, in the sense that the aviator can walk away from them unhurt, but there are no minor repairs to an airplane. They must be done by certified mechanics and have the O.K.

of the Civil Aeronautical authorities. Since they generally occur at remote points, they constitute a real burden to the pocket.

For many years, Bigelow nursed a series of old and decrepit cars. For the ordinary automobilist a car is an instrument to get somewhere, but for the enthusiastic gadget man it represents a challenge to his ability to overcome difficulties. Such an engineer will never be content with a car that functions normally. He is either trying to construct a super car or he is exercising his ingenuity in making run a car which, by all the canons of the motorist, should have been consigned to the junk heap years ago. If you ride with such a motorist, you will be safe from accidents of any consequence, but you will never, never travel without adventure. I remember one occasion when Von Neumann was interested in consulting with Bigelow as a possible engineer for a computing machine project. We telephoned from Princeton to New York, and Bigelow agreed to come down in his car. We waited till the appointed hour and no Bigelow was there. He hadn't come an hour later. Just as we were about to give up hope, we heard the puffing of a very decrepit vehicle. It was on the last possible explosion of a cylinder that he finally turned up with a car that would have died months ago in the hands of anything but so competent an engineer.

Bigelow and I began to try to ascertain the limitations of our prediction method, for it was almost certain that we should find serious limitations. This time, instead of trying out our predictor on a smooth curve, we tried it on graphs made of two straight lines joining each other at an angle.

It must be understood that the predictor consisted of one member which was made to follow a given curve and of another member which, on the basis of these past data, was expected to indicate the curve a little further in the future. This second member we will term the follower. When we put our apparatus to work on a curve which was not smooth, but in which one straight segment of a line was succeeded by another segment at an angle with the first, the predictor still worked, but in a very peculiar fashion.

What was interesting and exciting, and in fact not unexpected, was that the pieces of apparatus designed for best following a smooth curve were oversensitive and were driven into violent oscillation by a corner. We tried to test this repeatedly, and we always got the same result. Then the idea suggested itself to me: perhaps this difficulty is in the order of things, and there is no way in which I can overcome it. Perhaps it belongs to the nature of prediction that an accurate apparatus for smooth curves is an excessively sensitive apparatus for rough curves. Perhaps we have here the example of the same sort of malice of nature which appears in Heisenberg's principle, which forbids us to say precisely and simultaneously both where a particle is and how fast it is going.

The more we studied the problem, the more we became convinced that we were right and that the difficulty was fundamental. If then we could not do what we had wanted but had not really hoped to do (that is, to develop a perfect universal predictor), we should have to cut our clothes to fit our cloth and develop the best predictor that mathematics allowed us to. The only question was: What did we mean by the best predictor? If errors of inaccuracy and errors of hypersensitivity always seemed to be in opposite directions, on what basis could we make a compromise between those two errors?

The answer was that we could make such a compromise only on a statistical basis. For the actual distribution of curves which we wanted to predict, or let us say for the actual distribution of airplanes that we wanted to shoot down, we might seek a prediction making some quantity a minimum; and the most natural quantity to choose at the start, if we should be guided by easy computation, if not military significance, was the mean square error of prediction.

This means that we took the square of the error of prediction at each time, or in other words, the square of the difference between the predicted value and the true value. We then took the average of this over the whole time of the running of the apparatus. This average of the square error was what we were trying to minimize.

Thus we could set up the prediction problem as a minimization problem and give it a definite mathematical form once we made certain assumptions concerning the statistics of curves to be predicted. The branch of mathematics dealing with the minimization of quantities associated with curves is known as the calculus of variations, and it has a very well-known and well-recognized technique. In many cases it leads to setting up a certain differential equation for the function or curve which is to fulfill the minimization condition, but there are cases (and this case of ours was one of them) where it leads to the related sort of equation known as an integral equation.

This was lucky for me, for integral .equations were well within my field of interest; but the even luckier thing was that the particular integral equation to which the problem leads is a slight extension of the one which had been considered by Eberhard Hopf and myself. The result was that not only was I able to formulate the prediction problem but also to solve it; and what was even luckier was the fact that the solution came out in a simple form. It was not hard to devise apparatus to realize in the metal what we had figured out on paper. All that we had to do was make a quite simple assembly of electric inductances, voltage resistances, and capacities, acting on a small electric motor of the sort which you can buy from any instrument company.

We made an apparatus which translated the height of a point above a given base line into an electrical voltage. We passed this voltage, which varied with the time, through an electrical combination of resistance wires, condensers, and magnetic coils. At another point in the system, we took off the voltage and measured it continuously by a voltmeter. The actual type of voltmeter which we used gave us a continuous graph of the output voltage. It was this output which was to serve as a prediction of the voltage a certain length of time in the future.

The next problem which I had to take up concerned prediction in the case in which the data from which we were to predict were not precisely given. This also led to a minimization problem, in which we had to specify not only the statistics of the data which we were supplied but also the statistics of the errors at the same time. This minimization problem led to another Hopf-Wiener equation. This was solvable by the same methods, and we obtained a very satisfactory theory.

In scientific work it is not enough to be able to solve one's problems. One must also turn these problems around and find out what problems one has solved. It is frequently the case that, in solving a problem, one has automatically given the answer to another, which one has not even considered in the same connection.

This proved to be true in the new prediction theory. The concept of predicting the future of a message with a disturbing noise on the basis of the simultaneous statistics of the noise and message turned out to contain in itself the whole idea of a new method for separating noises and messages in what would be in some sense the best possible way.

This happened at a very opportune time, for the new technique of radar had come to meet serious difficulties. In radar too it was important to pick out a confused and faint message from the background of noise. Noise to the electrical engineer means not only noise that you hear but any unwanted electrical disturbance. For example, the flutter and flicker which you see in a badly regulated television set is noise. The messages which come in through a piece of radar apparatus and serve to confuse the image for which we are looking instead of to define it are known as noise.

The separation of noise from messages is the function of a wave filter. Wave filters go back many years in the history of telephone engineering and are pieces of apparatus which manage to free a message from part of the accompanying noise. Originally they were designed to pass all messages over a certain range of frequencies (or pitches) with as little change in their intensity as possible and to weaken other immediately adjacent frequencies by as much as possible.

When, with the precedent of telephone filters, such filters were built for television, and it was found that, after a certain point, the sharper the design of the filter, the worse it would function. Why was this the case? The answer is that the telephone filter is tailored to the specific characteristics of the human ear. The human ear is a very accurate instrument for perceiving pitch and a fairly accurate one for perceiving loudness, but it is a very poor instrument for perceiving what is known as phase; or, in other words, the precise time that the oscillation of the air passes through zero. An alternating current is represented, as I have said, not by one quantity which gives it intensity, but by two, which give it intensity and phase. The picture of an alternating current somewhat resembles the teeth of a comb. As I move the comb forward and backward along its edge, I change a certain quantity known as phase. In sounds, this phase change is not completely imperceptible, but it is not particularly important, and the earlier filters for telephone work and for other sound work did not pay much attention to differences in phase.

Radar, like television, appeals to the eye; and in the sort

of message that radar or television transmits, the eye is quite as sensitive to phase errors as to amplitude errors. Thus, the phase distortion which the old telephone-type filter generated in radar and television was too high a price to pay for the excellent transmission of amplitude over a large range of frequencies. To minimize the total error in television and radar, it was necessary to cut down the phase distortion, at the cost of permitting a little more amplitude distortion than would by itself be best. In seeking for this balance between the two distortions, the method which I had suggested—although not ideal—would at least work, and was far better than any which had previously been used.

I do not mean to say that others had not become aware of the failure of the earlier forms of filter design for radar nor that these other people had not understood what was in essence the reason for this failure, but simply that my method gave for the first time a simple, compact, and reasonable way of attacking the problem on the level of fundamentals.

Bigelow and I started up a little laboratory to explore the possibilities of predictors. We had a staff of two men working with us. One was an excellent machinist and electrician, who put our ideas into the metal almost as fast as we could conceive them. The other was a computer who had been an accountant.

Reader, if you ever have to start a computing laboratory, be warned by me and do not take as a computer an accountant, no matter how honest and efficient. Your computer must work to so and so many places of accuracy. This means so and so many significant figures, whether the significance of the digits begins six places before or six places after the decimal point. Your accountant works to cents, and he will work to cents until hell freezes over. Whatever numbers our accountant computed he kept at all stages to exactly two places after the decimal point, whether they were numbers in the millions, where even the first place to the left of the decimal point was of no possible significance, or numbers which begin only five places after the decimal point.

This was his conscience, that he should be accurate to the last cent; and he simply could not understand that physical quantities are not measured in cents but on a sliding scale of values in which the cents of one problem might be the dollars of another. In particular, when he had to obtain a small number as the difference between two nearly equal large numbers, he could never realize that these large numbers would have to be measured to a much greater degree of accuracy than that which was to be demanded for their difference.

I took my responsibility in this project very seriously. I tried to work against time, and that is a thing for which I am completely unsuited. More than once I computed all through the night to meet some imaginary deadline which wasn't there. I was not fully aware of the dangers of Benzedrine, and I am afraid that I used it to the serious detriment of my health.

Be that as it may, I found one very disagreeable fact, that the burden of secrecy in my project weighed heavily on me and that Benzedrine plays hob with one's ability to keep a secret properly. It superimposed on my not very secretive nature a garrulity which was completely unfitting at the time. I had to give it up and look for a more rational way of strengthening myself to bear the burdens of war work.

My work was under the supervision of Dr. Warren Weaver, of the Rockefeller Institute. We made several trips, Bigelow and I, to consult with him, and to compare our ideas with those of other people working on prediction theory and on the smoothing of anti-aircraft data. We traveled south two or three times to Fort Monroe, in Virginia, and to an army camp on the North Carolina coast. There we found workers from the Bell Telephone Laboratory who were more than eager to exchange ideas with us, and we pooled our resources with them and with other workers in the same field. At these meetings, after our travel and our hard work, I am afraid that I often went to sleep.

When we got back home, we made an experimental setup for generating the sort of irregular functions which arise in the aircraft-prediction problem, and then we designed a prediction apparatus on the basis of the statistical observations which we made on this setup. We actually were able to construct a predictor which would show the shape of a pattern of voltage in time, let us say, half a second before it occurred. This allowed us to check on our theory and find the criterion for a piece of apparatus that would give us good prediction.

The problem of generating an irregular curve with a statistically controllable degree of irregularity was quite interesting. We had reflected on the ceiling a spot of light, moving in a more or less periodic course. We tried to follow this spot with another, reflected in a mirror which was controlled by a certain apparatus. In this apparatus, the actual motion of the spot was not proportional to the turning of the crank which regulated it but to a rather complicated mixture of derivatives and integrals of this motion. Moreover, the crank was attached to a system of weights and springs, which was very far from giving the kinesthetic sensations which one would naturally associate with an apparatus of the sort. In other words, the spot had to be moved by a control which was complicated to begin with and, furthermore, felt completely wrong. Naturally, each person would respond to the apparatus in a somewhat different way; and we based our set of the predictor not merely on the general behavior of the apparatus but on the specific ability of an individual person to control the apparatus at a certain specific epoch in his training.

We were gratified by our clear and consistent results. On one hand, we had made a mechanical setup which threw a great deal of light on the way in which we act when we are confronted with an artificial problem and on the nature of humanly caused irregular action. On the other hand, we had found a way to duplicate in some degree the properties of the type of irregular motion of an airplane in flight. Thus we had some hope of the theory which could be used for the design of a practical apparatus for bringing down airplanes.

The importance of our ideas in connection with the control of anti-aircraft fire was double. There are two human elements which must be considered in this control. On the one hand, when the airplane pilot is flying and taking evasive action, his pattern of flight has a great deal to do with not only the limitations of his plane but those of his nervous system, so that his action is not too different from that of the hypothetical human action we had designed. On the other hand, the anti-aircraft gunner uses a technique in which he cannot follow his target perfectly but in which he introduces certain random errors because of the limitations of his sense organs and muscles. These two sorts of human elements are combined as part of the semimechanical processes by which the anti-aircraft gunner brings down his target.

At the beginning of the war the only known method of tracking an airplane with an anti-aircraft gun was for the gunner to hold it in his sights by a humanly regulated process. Later on in the war, as radar became perfected, this process was mechanized. It became possible to couple directly to the gun the radar apparatus by which the plane is localized, and thus to eliminate the human element in gun pointing.

However, it does not seem even remotely possible to eliminate the human element as far as it shows itself in enemy behavior. Therefore, in order to obtain as complete a mathematical treatment as possible of the over-all control problem, it is necessary to assimilate the different parts of the system to a single basis, either human or mechanical. Since our understanding of the mechanical elements of gun pointing appeared to us to be far ahead of our psychological understanding, we chose to try to find a mechanical analogue of the gun pointer and the airplane pilot.

In both these cases, the operators seemed to regulate their conduct by observing the errors committed in a certain pattern of behavior and by opposing these errors by actions deliberately tending to reduce them. This method of control appeared to us not unlike a method already known in electric circuits and now being applied in servomechanisms, or systems by which we switch in an outside source of power for control purposes, such as occurs in the power steering of a truck. We call this negative feedback.

We use negative feedback in controlling the power input to the gun turret of a ship. When the direction in which the gun is pointing and the direction in which our computing apparatus says that the gun ought to point are different, this difference is used to regulate a power input to the turret, which will be of such a nature as to bring the turret more nearly into the desired position.

It is a maxim of the physiologist that the pathology of an organ throws a very great light on its normal behavior. We asked ourselves the question: Does a negative-feedback apparatus have a specific recognizable pathology? Here we were on firm ground.

To see the general purpose of negative feedback apparatus, let us take the case of a gun turret controlled by a handle. If this handle works the gun turret directly, then the same pressure on the handle will produce very different results if the gun turret is cold and the grease is sticky from what it will produce if the gun turret is warm and the grease flows easily. It will produce different results if the gun is depressed, thereby increasing the moment of inertia of the gun turret, from those which we shall find if the gun is elevated and the turret has a small moment of inertia about a vertical axis. The primary purpose of feedback control on the gun turret is to make the response of the turret more nearly proportional to the push on the lever and thus less dependent on the variable friction, the inertia, and other external circumstances.

Not only is a feedback system less dependent on changes of load than a system without feedback, but this dependence becomes increasingly less as more and more of the motion is fed back—or, in other words, if the feedback is put through larger and larger amplification. However, this improvement in behavior does not continue indefinitely, for after a certain stage, with a large measure of amplification in the feedback, the apparatus will go into spontaneous oscillation and behave in such a wild way that we have decreased rather than increased the load independence of the apparatus. We expected that if human control also were to depend on feedback, there would be certain pathological conditions of very great feedback, under which the human system, instead of acting effectively as a control system, would go into wilder and wilder oscillations until it should break down or at least until its fundamental method of behavior should be greatly changed.

This suggestion, which emanated equally from Bigelow and from me, I brought to the attention of my neurophysiologist friend, Dr. Rosenblueth. He had not yet left for Mexico and was still Dr. Cannon's colleague at the Harvard Medical School. The specific question we put was: Are there any known nervous disorders in which the patient shows no tremor at rest, but in which the attempt to perform such an act as picking up a glass of water makes him swing wider and wider until the performance is frustrated, and (for example) the water is spilled?

Dr. Rosenblueth's answer was that such pathological conditions are well known, and are termed intention tremors; and that very often the seat of the disorder lies in the cerebellum, which controls our organized muscular activity and the level on which it takes place. Thus, our suspicions that feedback plays a large role in human control were confirmed by the well-established fact that the pathology of feedback bears a close resemblance to a recognized form of the pathology of orderly and organized human behavior.

Within the last two years I have had an experience which may be regarded as a commentary on the ideas I am putting forward here. Suddenly my little granddaughter, who was staying with us, developed a purpose tremor of exactly the nature of the one discussed here. We took her at once to the hospital and found that she was suffering from some form of encephalitis involving the cerebellum. This was a condition with very grim possibilities, but she had the splendid luck to make a perfect recovery without any aftereffects. If I were a superstitious man, this experience and many others reported by medical men might make me suppose that a disease has a vicious personality and wished nothing more than to revenge itself on the scientist who has pursued it.

To go back to the work of our team of three: we wrote up these ideas in an article, but what was even more important was that Bigelow and I felt that we could safely go ahead with the treatment of the human links in the control chain as if they were pieces of feedback apparatus. Accordingly, we felt justified in proceeding from our crude experimental setup in the direction of the design of a complete apparatus for anti-aircraft control and prediction.

Since it was clear that the anti-aircraft control apparatus was in essence a feedback loop and contained in its construction many subsidiary feedback loops, we had to find out something of the characteristics of these loops. These characteristics were not available, so that the over-all apparatus which we designed was essentially crude and of unverifiable behavior. Under the circumstances, it was not considered advisable to proceed much further with it, partly as such

tentative behavior as we could compute mathematically did
not suggest an excellent performance.

Our ideas were eagerly taken up by other workers in the
field and did lead to a very definite improvement in practice,
in particular of the part that consisted in filtering out experimental errors of observation. We were not finally commissioned
to perfect our own design, but I was in fact asked to write a
book on time series, extrapolation, and interpolation. This book
was reproduced in a photo-lithographed form which, because
of the yellow paper in which it was bound, came to be known
as the "yellow peril," a name previously confined to the yellow
books of Springer's mathematical series. My textbook was very
freely used, not only during the war by the designers of systems
of control for the aiming and firing of anti-aircraft guns, but
also by servo-engineers and electrical-communication men,
both then and later. It was reprinted after the war in an amplified and improved edition with an appendix by Professor
Norman Levinson, who made the use of my methods considerably clearer.

The work I did on the statistical treatment of anti-aircraft-
fire control has led eventually to a general statistical point of
view in communication engineering. The years that have gone
by between then and now have secured the general acceptance
of this point of view in communication engineering, but they
have also led to much more. I might almost say that the
whole of engineering is rapidly assuming a statistical aspect,
and that this is passing over to less orthodox fields such as
meteorology, sociology, and economics.

Let me go back to my earlier remarks concerning Willard
Gibbs and the revolution he and his contemporaries created
in physics. The orthodox Newtonian view of physical dynamics
provides for certain equations involving rates, known as differential equations. With the aid of these equations and with
a knowledge of the initial values, or values at time zero, of the

variables whose rates are being determined in relation to their numerical values, we can inch our way along in developing the history of these phenomena. At each time we know our values. From these we determine their own rates of change, and this gives us an approximation to the same values and rates an instant later.

If we choose an instant of time that is sufficiently short, we may progress along the history of our phenomena to any point in time which we wish to attain. This is the method of the astronomers in calculating the orbits of the planets and the method of the ballistics expert for the study of the flight of projectiles in determining their paths.

In astronomy, as I have said before, the computation of these orbits is a very precise mathematics, and our initial data are very accurately known. This is not the case in most ballistic or engineering problems. In firing a shell, for example, the angle of elevation is known only to a very limited degree of accuracy, as are the weight of the projectile, the charge of powder, and the various atmospheric conditions. The result is that we start with none of this precisely given but with each set of data given only within a certain range. The traditional way of solving the ballistic equation is to assume the initial data as given precisely. Then we find the range, the angle of impact, the velocity of impact, and other significant quantities, and we immediately start to revise these with the aid of methods of interpolation or correction, reckoned by procedure which is entirely distinct from the first.

In this process we waste a good deal of effort, first in making our data unrealistically accurate, and second in correcting our imperfectly realistic results. There is, however, another method which is now coming into use, and which finds its spiritual father in Willard Gibbs.

Gibbs pointed out that when a dynamical system develops according to its own laws as, for example, when a top spins freely, something occurs very much like the flow of a fluid. To characterize a top, we need a point in a certain space, but this space is not the same as the familiar three-dimensional one of solid geometry. The position of a top requires six co-ordinates—or measurable quantities—to give its position and six to give its momenta; and these together form a twelve-dimensional array, which we may call by analogy a twelve-dimensional space. In this space there is a certain measure of volume, so that a set of tops which fill a given volume at one time will fill an equal volume at any other. This type of invariance of volume is to be found in all dynamical systems in which there is no power input or output.

This flow may be conceived as a flow of probability, and so it was conceived by Gibbs. The probability that a particle will be at one time in a given region of this peculiar space is the same as that at a later time it will be in the corresponding region into which the first has flowed.

Thus the typical equation of flow is no longer a general system of what is known as ordinary differential equations but a set of integral equations. These integral equations relate past distributions to future distributions, in such a way that if we superimpose different past distributions we superimpose the corresponding future distributions. A system like this, in which sums in the output answer to sums in the input, is known as linear, and the integral equations of the flow method of treating dynamics may be taken as linear.

The method is quite practicable computationally; and if the problem to be solved is of any high degree of complexity, this method may well be easier than the purely Newtonian method. Certain simplifications of methods of this type are now being used extensively by members of the mechanical-engineering department at M.I.T.

In addition to purely computational advantages in the more complicated cases, this method is also essentially superior to the Newtonian method of computation from the logical point of view. The reason is this: what we put into our problem not only consists of precise data which we later have to ease off in accordance with the inaccuracy of the equations and the initial conditions but contains intrinsically the very inaccuracy which hinders our work. We are thus not overcomputing and relieving the effect of this overcomputing by an *ad hoc* study of its errors but putting all our cards on the table at the beginning. What we finally get is what we want, neither more nor less. This cuts down a lot of unnecessary effort, but it also increases the real precision of what we are doing.

No scientific measurement can be expected to be completely accurate, nor can the results of any computation with inaccurate data be taken as precise. The traditional Newtonian physics takes inaccurate observations, gives them an accuracy which does not exist, computes the results to which they should lead, and then eases off the precision of these results on the basis of the inaccuracy of the original data. The modern attitude in physics departs from that of Newton in that it works with inaccurate data at the exact level of precision with which they will be observed and tries to compute the imperfectly accurate results without going through any stage at which the data are assumed to be perfectly known.

If we follow in these unprecise problems the sort of computation which the astronomer uses in determining the orbits of the planets, we may happen to choose initial conditions which lead to final results not typical of the wider range of initial conditions with which we have operated, and this instability of our orbit may drive us to a false reckoning of our ultimate error.

As I have said earlier in the description of my work on prediction, the more sensitive our instruments become, the more unstable they will be. These cause an error of a different sort from that of imprecision, but an error equally serious. What

I have said of mechanical instruments is true of methods of computation. The balance between errors of imprecision and errors of instability is something which we can compute only on a statistical basis. Why not, then, assume the statistical basis at the very beginning and obtain both mean result and error by a unified method of computation?

If this recognition of the statistical nature of all science is already proving to be valuable in the most Newtonian type of mechanical-engineering computation, how much more must it then be the natural method of computation in those fields in which our errors of observation are naturally very large!

Let us consider meteorology as one example. We know a good deal about the dynamics of the atmosphere; and if our observations of the initial conditions were extremely good, we might expect that we could compute the future in a purely Newtonian way, even though this way is very likely to involve a great deal of overcomputation. What we actually know about the atmosphere, though, is a sampling taken in no more than three or four observations per day per hundred thousand cubic miles of the atmosphere.

Recently, under the influence of John von Neumann, there has been an attempt to solve the problem of predicting the weather by treating it as something like an astronomical-orbit problem of great complexity. The idea is to put all the initial data on a super computing machine and, by a use of the laws of motion and the equations of hydrodynamics, to grind out the weather for a considerable time in the future.

The catch is that all the observations of the weather bureau give only limited information at a very few points, with colossal gaps between them. These one can fill up only by some sort of statistical reasoning. Thus, an adequate meterological method must partake both of dynamics and statistics. There are clear signs that the statistical element in meterology cannot be minimized except at the peril of the entire investigation.

I do not mean to deny the importance of dynamics, but I do mean to assert the virtues of that Gibbsian approach in which this dynamics is treated as a statistical flow.

Meterology is typical of most of those numerical sciences which have come to the fore late in the history of science. In economics, the so-called econometric science of economic dynamics suffers under the radical difficulty that the numerical quantities which are put into the dynamics are not well defined and must be treated as gross statistical estimates. Who knows precisely how to define a demand, and how to measure it in terms which will satisfy most other economists? Can any two economists check on the amount of unemployment in the United States at a given time?

Econometrics will never get very far until two steps are taken: One of these steps is that the observation of the quantities—demands, inventories, and the like—with which econometrics operates must be subject to the same criteria of precision and rigor as the dynamics by which they are combined. The other is that we should recognize from the beginning the statistical and imperfectly precise nature of the quantities with which we operate and that we should go over to a Gibbsian treatment of them.

What I have just said about meteorology and econometrics applies equally to sociological dynamics, to biometrics and in particular to the very complicated study of the nervous system which is itself a sort of cerebral meteorology. It belongs to the very grammar of the use of mathematical methods in semi-precise sciences. It is the heart of the engineering of the future.

This new technique was foreshadowed in my war work on anti-aircraft fire-control predictors and was carried further in my development of communication theory. As yet it has penetrated to only a few initiates in the appropriate fields of scientific work, but it is philosophically right, and it bids fair to change the entire face of all the precise and semiprecise sciences.

When I first wrote about prediction theory, I was not aware that some of the main mathematical ideas had already been introduced in the literature. It was not long before I found out that just before the Second World War an important little paper on the same subject had been published by the Russian mathematician Kolmogoroff in the *Comptes Rendus* of the French Academy of Sciences. In this, Kolmogoroff confined himself to discrete prediction, while I worked on prediction in a continuous time; Kolmogoroff did not discuss filters, or indeed anything concerning electrical engineering technique; and he had not given any way of realizing his predictors in the metal, or of applying them to anti-aircraft-fire control.

Nevertheless, all my really deep ideas were in Kolmogoroff's work before they were in my own, although it took me some time to become aware of this. A series of papers by Kolmogoroff and such pupils of his as Krein continued to appear in the *Doklady (Reports of the Russian Academy of Sciences),* and although these papers still stuck for the most part to the concept of prediction previously developed by Kolmogoroff, somewhat narrower than my own, I am by no means convinced that Kolmogoroff was not independently aware of the possibility of some of the applications I had made. If that was so, he must have had to keep them out of general publication because of their importance for the military-scientific work of the Soviets. A recent paper by Krein, in which he makes an explicit allusion to my own work in the field of applications, convinces me of this.

I have never met Kolmogoroff, and indeed I have never been in Russia, nor have I been in correspondence with him or with any of his school. Thus what I say about him is largely surmise. At an early stage of my work for the United States military authorities, before I had seen Kolmogoroff's paper, the question came up whether anybody abroad was likely to be in possession of ideas similar to mine. I said that they would unquestionably receive no particularly ready reception in Germany; that my own friends Cramer, in Sweden, and Lévy, in France, might well have been thinking along similar lines; but that if anyone in the world were working on these ideas it would most likely be Kolmogoroff in Russia. This I said because of my knowledge that for twenty or thirty years hardly had either of us ever published a paper on any subject but the other was ready to publish a closely related paper on the same theme.

Within the last two or three years I have seen a Russian book on prediction theory, communication theory, and similar topics which makes extensive references to both Kolmogoroff's work and to my own. It gives Kolmogoroff the priority, and although this priority is only partial, I have just said that there is good reason for considering him not only as an independent discoverer of large parts of the subject but as the first man to write on it. The book takes my own work very seriously and treats me much more fairly than I would expect in a Soviet book, international relations being what they are.

The "yellow peril" book is still playing an important role in American research work, both for military purposes and for more general uses. It was with the permission of the government that it has been reprinted, and it must have been a copy of this book which filtered into Russia and served as a basis for the Russian comments of which I have just written.

From this point on, my work, or rather the work of my group, has spread out to cover a very wide field of communication theory and practice. In the first place, the "yellow peril" is most definitely a statistical treatment of problems of communication. When the book was written, almost nobody had thought of communication in these terms. I think I am to be pardoned for a certain pride in saying that the statistical approach to communication theory is now accepted almost everywhere.

I approached information theory from the point of departure of the electric circuit carrying a continuous current, or at least something which could be interpreted as a continuous current. At the same time, Claude Shannon, of the Bell Telephone Laboratories, was developing a parallel and largely equivalent theory from the point of view of electrical switching systems. This represented a direct development of his previous work on the use of the algebra of logic in switching problems.

As I have said before, Shannon loves the discrete and eschews the continuum. He considered discrete messages as something like a sequence of yeses and noes distributed in time, and he regarded single decisions between yes and no as the element of information. In the continuous theory of filtering, I had been led to a very similar definition of the unit of information, from what was at the beginning a considerably different point of view.

In introducing the Shannon-Wiener definition of quantity of information (for it belongs to the two of us equally), we made a radical departure from the existing state of the subject. For many years it was believed that the carrying power of a communication line per unit time was to be measured by the band width it could carry.

A band width of 200 cycles was supposed to be able to carry twice as much information per second as a band width of 100 cycles. This supposition ignored the fact that, in the absence of noise, any band width would be enough to carry any amount of information in one second. One single voltage measured to an accuracy of one part in ten trillion could convey all the information in the Encyclopedia Britannica, if the circuit noise did not hold us to an accuracy of measurement of perhaps one part in ten thousand.

In the early days of the telephone art, very few lines were

burdened with messages to the ultimate limit of their message-carrying capacity. As the art developed and the new modes of communication like radio and television demanded a more complete exploitation of the message space available, it became clear that the noisiness of the line or air channel is another important factor which we must take into consideration. The ether is full of disturbances which the radio man terms static, and no conductor, be it metallic or gaseous, can carry electricity in smaller lumps than the single electron. The irregularity of the stream of electrons is known as the shot effect, and it is an important consideration in all modern communication design.

It was only shortly before World War II that the load on communication channels became heavy enough for this intrinsic noise to become a serious practical barrier to the use of the lines for even more communication. Thus, the statistical point of view in communication theory, which I had anticipated so many years before with my generalized harmonic analysis, and which Shannon and I had jointly made fundamental at the beginning of World War II, became inevitable and basic shortly after the war had begun.

The work we were doing on feedback in connection with the fire-control machine and the nervous system introduced another revolution, which, like the first, has received universal acceptance in the course of the last few years. When I first came to Tech, electrical engineering was divided into two fundamental parts, which were known in Germany as the technique of weak currents and the technique of strong currents, and in the United States as communication engineering and power engineering.

The distinction between these two fields is valid, but the nature of this distinction and the place to make it was not understood for a considerable period. The generating station for a television sender or a trans-Atlantic radio sender may use relatively large quantities of power, but it is directed primarily towards communication; while the fractional-horsepower motor used in a dentist's drill may employ relatively small quantities of power. Nevertheless, the first piece of apparatus is primarily oriented with respect to the message and the second one with respect to the energy consumed.

At a period at which this distinction was not fully appreciated, servomechanisms for the control of gun turrets and other pieces of heavy apparatus were naturally assumed to belong to power technique rather than to communication technique. The whole tradition of power technique was to consider electric currents and voltages as varying in time, while the whole tradition of communication technique, particularly under the influence of Heaviside, had led to consideration of a message as a sum of a large or infinite number of different frequencies. It was not easy to see that the frequency treatment, rather than the time treatment, was just as appropriate for the servomechanism as for the telephone, the telegraph, and television.

I think that I can claim credit for pointing this fact out and for transferring the whole theory of the servomechanism bodily to communication engineering. My whole point of view in these matters made me regard the computing machine as another form of communication apparatus, concerned more with messages than with power. Its nature, as I saw it, was that of a series of switching devices, so enchained together that the information coming out of a number of stages of these was introduced into a subsequent stage as ingoing and regulating information.

It was clear that while these switching devices might be gear wheels and the like, they could equally well be mechanical relays or the electrical relays which depend on vacuum tubes and other electronic phenomena. I was much more disposed, as I have said, to use switching devices which made a choice

between two alternatives than those which made a choice between ten alternatives, and I tried to bring this concept of
computing machines to the attention of the engineering public.

It was at Harvard, under the supervision of Howard Aiken,
that I found the first of the newer switching computers dependent on relays. Aiken was developing them under a government grant. I was much struck with Aiken's work, which I
greatly admired, and which Aiken himself considered as the
modern carrying-out of the crude computers developed by
Babbage in England a hundred years ago. Babbage had formed
an excellent conception of their mathematical possibilities but
had almost no understanding of the mechanical problems to
which they gave rise.

I was surprised to find that Aiken was completely committed
to the relatively slow mechanical relay as the mechanical computer's first tool and that he did not put any enormous value on
the speed which could be derived by the use of electronic
relays. This limitation of point of view has now been discarded
by Aiken himself, who has become one of the most active and
original inventors and designers of electronic computers. But
at that time he labored under a curious moralistic quirk in
accordance with which he considered work with mechanical
relays as essentially sound and right and work with electronic
relays as unnecessary and ethically sloppy.

Here I wish to emphasize again a weakness of attitude joined
with a great strength in those men who show practical ingenuity in the devising of gadgets. It is the desire to fix the
technique of a subject forever at the precise point to which
their ingenuity has carried them and then to offer a profound
intellectual and moral resistance—a block, in fact, to later work
which departs from their principles. We mathematicians who
operate with nothing more expensive than paper and possibly
printer's ink are quite reconciled to the fact that, if we are
working in a very active field, our discoveries will commence

to be obsolete at the moment they are written down or even at the moment they are conceived. We know that for a long time everything we do will be nothing more than the jumping-off point for those who have the advantage of already being aware of our ultimate results. This is the meaning of the famous apothegm of Newton, when he said, "If I have seen further than other men, it is because I have stood on the shoulders of giants."

Yet the commercial possibilities of the invention in the metal tend to blind the industrial worker to this fundamental fact and to make him hope that he can hold back the stream of progress at the precise stage where he had made his own contributions. The patent system and the commercial value of an inventor's idea as something salable tend to push him in this direction. This is not realistic. As a practical man, the inventor should have the very practical consciousness that for many years his greatest contribution will not be a single gadget but the furthering of the whole stream of thought and ideas concerning an enormous class of gadgets past, present, and future. He should come to terms with this streaming of thought and realize that, just as he has gone beyond those who were born before him, he himself and his work will have to serve rather as a stepping stone to the future than as the end to which science and technique must finally arrive.

However, my interest in the development of computing machines carried me far beyond those machines past, present, or to come, which are made of brass and copper, glass and steel. The brain and the nervous system also share in the main characteristics of computing machines. Parallel to the yes and no of a relay is the fact that a nervous fiber can exist in what are fundamentally only two states: the state of carrying a message and the state of not carrying a message. This is the so-called all-or-none law of the nervous system; and, although it may not be as precisely true as its crude, cold formulation would suggest, it is sufficiently true to represent a fundamental fact of nervous conduction.

A nerve fiber, it is true, may be stimulated by messages of varying intensity, but the ultimate fate of each of these messages is either to die out and fail to reach the end of the fiber or to continue as what the chemists would call a self-catalyzing process and start an impulse which will go from one end of the fiber to the other. When it has reached the end of the fiber, its subsequent history is so nearly independent of the original strength of stimulation that this strength may be entirely neglected. Thus, there is a certain analogy between a nerve fiber and a flip-flop electric circuit, an electric circuit with two, and only two, states of equilibrium. This analogy is so close that, long before the message reaches the end of the fiber, it carries its information in the form of a number of impulses rather than in the form of the strength of the impulses.

Not only are nerve fibers switching devices, but they are devices which lead into other switching devices. The nerve fibers communicate with one another by junction points or junction systems known as synapses, and in these the question whether a new message is established in an outgoing fiber depends on the precise set of incoming messages received from various fibers. In the simplest cases, the synaptic system has a threshold, which means that if more than a certain number of incoming fibers receive messages within a certain critical interval of time, the outgoing fiber fires, and otherwise not.

We are so used to feedback phenomena in our daily life that we often forget the feedback nature of the simplest processes. When we stand erect, it is not in the manner in which a statue stands erect, because even the most stable statue needs to be fastened to some sort of pedestal or it would fall over. Human beings stand erect, however, because they are continually resisting the tendency to fall down, either forward or backward, and manage to offset either tendency by a contraction of muscles pulling them in the opposite direction. The equilibrium of the human body, like most equilibria which we find in life processes, is not static but results from a continuous interplay of processes which resist in an active way any tendency for them to lead to a breakdown. Our standing and our walking are thus a continual jujitsu against gravity, as life is a perpetual wrestling match with death.

In view of this, I was compelled to regard the nervous system in much the same light as a computing machine, and I communicated this idea to my friend Rosenblueth and to other neurophysiologists. I managed to get a group of neurophysiologists, communication engineers, and computing-machine men together at Princeton for an informal session, and I found on the part of each group a great willingness to learn what the other groups were doing and to make use of their terminology. The result was that very shortly we found that people working in all these fields were beginning to talk the same language, with a vocabulary containing expressions from the communication engineer, the servomechanism man, the computing-machine man, and the neurophysiologist.

For example, all of them were interested in the storage of information to be used later, and all of them found that the word *memory* (as used by the neurophysiologist and the psychologist) was a convenient term to cover the whole scope of these different fields. All of them found that the term *feedback*, which had come from the electronics engineer and was extending itself to the servomechanism man, was an appropriate way of describing phenomena in the living organism as well as in the machine. All of them found that it was convenient to measure information in terms of numbers of yeses or noes, and sooner or later they decided to term this unit of information the *bit*. This meeting I may consider the birthplace of the new science of cybernetics, or the theory of communication and control in the machine and in the living organism.

I had hopes that this new science was going to pass through a rapid development over a broad front. The subject has developed greatly, and I have participated in its later phases. However, the times were not favorable for the normal growth of new ideas, and I have had to watch very carefully through a period where what I intended as a serious contribution to science was interpreted by a considerable public as science fiction and as sensationalism.

Science fiction is in vogue, and it is the fashion even among certain serious scientists to see merit in its writings. I myself as a child was a devotee of Jules Verne and H. G. Wells, to whom the present literature of science fiction owes its origin, but it is an infinitely slicker and more pernicious article. On the one hand, it leads to fantasies of power and of brutality quite as devastating as anything in the thud-and-blunder type of gangster story or the most uncomic comics. On the other hand, it is helping to create a generation of youngsters who believe that they are thinking in scientific terms because they are using the language of science fiction. It is a real difficulty in our schools of science and engineering to have to try to educate young men who believe that they have a calling towards science merely because they are accustomed to playing with the ideas of destructive forces, other planets, and rocket travel.

This vicious daydreaming is largely a product of World War II, which has done so much to demoralize the whole generation of science. The period of the war was one in which the status of science and that of mathematics were changing rapidly. In the first place, leisure was vanishing in every sector of life. Before the war, I used to find the M.I.T. boys playing a game or two of bridge after lunch in one of the lounges of Walker Memorial. I often participated in these games.

I did not regard the time as wasted either by myself or by the students, for between games we used to have an occasion for wide-ranging discussions which might be pure bull session or might involve a real play of ideas. From the beginning of the war, everyone was in deadly earnest, and all chance of intellectual play was restricted. To the present day, it is hard to find young men who dare to take enough time off from work to consider what their work is about. The hours spent in the fantasy of space books are no replacement for a good bull session.

Before the war, and particularly during the depression, positions in science were not easy to get. The requirements for these positions had become exceedingly high. During the war, this situation had changed in two respects. First, there were not enough men to carry out all the scientific projects which the war involved. Secondly, in order to carry out these projects at all, it became necessary to organize the work so as to use those with a minimum amount of training, ability, and devotion.

The result was that young men who should have been thinking of preparing themselves in a long-time way for their careers lived in a lighthearted way from hand to mouth, confident that the existing boom in scientists had come to stay. Such men were in no state to accept the discipline or hard work, and they evaluated whatever intellectual promise they might have as if it had been already realized in performance. With the older men crying out for assistance and manpower, these boys would shop around for those masters who would demand least and grant them the most in indulgence and flattery.

This was a part of a general breakdown of the decencies in science which continues to the present day. In most previous times, the personnel of science had been seeded by the austerity of the work and the scantiness of the pickings. There is a passage in Tennyson's "Northern Farmer: New Style" which says: "Doänt thou marry for munny, but goä wheer munny is!"

Thus, an ambitious man with slightly anti-social tendencies

or, to put it more politely, indifferent to spending other people's money, would formerly have avoided a scientific career as if it were the plague itself. From the time of the war on, these adventurers, who would have started out as stock promoters or lights of the insurance business, have been invading science.

The old assumption which we used to make must be discarded. We all knew that the scientist had his vices. There were those among us who were pedants; there were those who drank; there were those who were overambitious for their reputations; but in the normal course of events we did not expect to meet in our world men who lied or men who intrigued.

When I began to emerge from my sheltered life into the scientific confusion of wartime, I found that among those I was trusting were some who could not be held to any trust. I was badly disillusioned more than once, and it hurt.

The meeting to consolidate communication theory took place well after Pearl Harbor. It may surprise the reader that in all this talk about war work, I have not mentioned Pearl Harbor and the actual entry of America into the war. The fact is that all of us had long been convinced that the war was coming to America in some form or other, and the actual opening of hostilities did not change my work on means of defense.

In the fall of 1941, the tension of the successive defeats experienced by the Allies in Scandinavia, Holland, and France, the Battle of Britain, and the ambiguous and to-and-fro situation in North Africa, had grown as great as one thought one could bear, yet it became complicated by the fairly general feeling that something was about to blow off in Japan. While none of us was exactly prepared for Pearl Harbor, I do not think that we were convinced that a military dictatorship like Japan would play the game of war and diplomacy according to the standard rules, particularly when these rules were manifestly advantageous to us. Thus, Pearl Harbor came, to me at least, as much more of a shame and a humiliation than a surprise.

Pearl Harbor and our subsequent entry into World War II on both sides of the globe had a number of direct effects upon me. It is true it could not involve me in war research any more than I was already involved, because I was fully occupied in this direction and had been for more than a year. However, the war wiped out plans which Manuel Sandoval Vallarta and I had made to go to South America in the interest of international good will, on funds primarily emanating from the State Department (or, in his case, from the Mexican Government).

What I felt to be much more important to me was what was going to happen to my dear friends the Lees. We had just managed to secure for them passage on a liner from Hong Kong. Then Pearl Harbor came and prevented the boat from sailing, or at least from sailing with our friends on board. Thus, Lee, who had already gone for five years through the Chinese-Japanese War without any adequate contact with his profession, was sentenced to wait five years more, until after V-J Day, before we could bring him over to the United States. During all that time, an offer which we had secured him from the electrical-engineering department of M.I.T. stood open, or at least ajar; and when he finally arrived he was able to step into an instructorship. This has now been succeeded by an assistant professorship and an associate professorship.

The problem of what professional man is to do when he comes back as a sort of Rip Van Winkle who has slept for a decade and wakes to find a changed world is very difficult. The obvious thing might appear to be to spend a year or two in studying the various developments of the intermediate period. This must be done to some degree or other, but it is not a completely adequate way of treating the situation. The very bulk of new material tends to produce an intellectual indigestion in the student. He must come into competition with the younger generation who have learned the field the easy way while it has been developing and who are at home in it. Our Rip Van Winkle cannot expect to compete with them.

What made Lee's situation easier was that I had recently developed a considerable part of the statistical theory of communication engineering in the "yellow peril." I pointed out to Lee that the one way to avoid being disastrously behind the game was to move deliberately ahead of the game and thus secure an advantage of some years while the other people were catching up. Lee saw the point.

The situation was made considerably easier by the fact that for years we had been so much in the habit of working together that my mental processes and ways of writing were quite familiar to him. Thus, he took over the problem of working out in detail the communication and engineering consequences of ideas which I had only sketched in general terms and of making himself interpreter to the engineering public of the field which I was later to call cybernetics.

Lee has established himself in this program and has been busy for some years in carrying it through to a most successful conclusion. He is now writing a book on communication engineering from the new standpoint, and he is showing in it a great patience, thoroughness, and consideration of the reader. For me, close as I am to the origin of the subject, such a detached treatment would be impossible.

Lee has presented the new ideas to quite a number of government and industrial laboratories. He has brought up a whole generation of young electrical engineers to do research along statistical lines and to employ my point of view as a habitual approach to communication problems. He has also organized very successful summer meetings, so that engineers already actively engaged in the communication industry have been able to come to M.I.T. for a refresher course on the cybernetic viewpoint.

In these ways the difficulties of ten years' isolation have been bypassed successfully. The head start which Lee has had in the new methods has allowed him time to catch up with what has happened between 1936 and 1946, and, what is more, the work in which he has been engaged has given him specific problems to use as a touchstone for his understanding of and familiarity with the research of the intermediate period. In other words, we have seen in the pay-off of the policy which the two of us began to adopt the moment the Lees arrived in Boston's South Station after the war.

With Lee back at M.I.T., I was greatly encouraged to go on with further investigations of servomechanisms and the entire class of topics which I was later to give the name of cybernetics. As I have said, Lee is himself now finishing a book on the matter. However, not all that the two of us could do together on that, nor all that any hundred people could do, would suffice to cover more than a small part of the literature on servomechanisms and on the automatic factory, to which our early work has led. The automatic factory bids fair to become the norm rather than the exception, even within the lifetime of those who are now in college. It is giving rise to a new profession of experts who are able not only to design these factories but to set up on them problems of the most varied sort. The modern technique of automatic-factory design is well beyond the ambit of a theoretical man like me. As I shall show in a later chapter, I have conceived it to be my primary function not so much to develop the automatic factory further as to explain its nature and its consequences, and to alert both labor and management to the need of facing these intelligently.

13

Mexico

1944

As the war went on, the work left the hands of the pure scientists like myself for those of the designers, and I was at loose ends. It was about January 1944 that I heard from Vallarta that there was to be a meeting of the Mexican Mathematical Society in Guadalajara the following spring and that they wished a good attendance from the United States, and my own presence in particular.

Almost from the moment I crossed the frontier, I was charmed by the pink-and-blue adobe houses, the bright keen air of the desert, by new plants and flowers, by the indications of a new way of living with more gusto in it than belongs to us inhibited North Americans. The high, cool climate of Mexico City, the vivid colors of the jacaranda blossoms and the bougainvillaeas, the Mediterranean architecture, all prepared me

for something new and exciting. The many times I have returned to Mexico have not belied these first impressions. It
will be a sad day for me when I come to feel that I have no
further chance to renew my contacts with that country and to
participate in its life.

The Rosenblueths met me when I arrived and saw that I was
put up in interns' quarters at the Instituto Nacional de Cardiología. As soon as I had begun to acclimatize myself and to
get over the terrible lassitude due to the elevation, Arturo and
I started to work together on the sort of muscle tremor known
as *clonus*: the familiar spasmodic vibration which many people
experience when they sit cross-legged with one knee under the
other. It seemed to be an admirable case for studying feedback
tremor in the neuromuscular system.

I had begun my trip with very little Spanish, what I had
being no more than a general knowledge of Latin and the
Romance languages and two or three lessons from a Mexican
student at M.I.T.

I brought a Spanish grammar along with me, and I tried to
make my very inadequate Spanish serve as a medium of communication with my interne companions, most of whom spoke
a very tolerable English. I used to dine with the interns in
the official doctors' dining room, and I found that I gradually
developed a taste for the strong pepper sauce, which seemed
to be the invariable accompaniment to all dishes. They were
very friendly to me and very courteous at the same time, for
both the Spanish language and the Mexican tradition permit
a rather close combination of courtesy and familiarity. Their
term for me was *maestro*, which is applied both to a teacher
and to a craftsman such as a carpenter or a mason. As applied
to a teacher, it is both far more familiar and far more respectful
than *señor*.

I played a lot of chess with the young doctors and occasionally with Arturo, although I think his estimate of me as a game

player was better indicated by the Chinese Checkers which we played continually when I visited him. I used to walk uptown and make purchases. For this I had to wait until my blood count had risen enough to overcome the functional anemia which besets all new visitors to Mexico City.

I learned a great deal more about Arturo during this visit. He had not started as a scientist but as a musician, and had earned his living for some time by playing classical piano music in a Mexico City restaurant. Arturo is also a first-rate chess player and a superb bridge player, so good that in neither of these games does he allow me to play with him. He is a great enthusiast for the climate and arts of his native land. In this I cannot gainsay him, although I think that he often tends to underrate the New England countryside and, for that matter, the country in general, as opposed to the city.

After Mexico, the land of which he speaks with the greatest admiration is France, where he did most of his medical studying. He would find quite enough pleasant things to say about New England if only he were speaking to a non-New Englander, but there is something puckish in him which makes him take a pleasure in teasing.

He is a hard worker, and he makes the greatest demands on the sincerity and industry of those about him, demands which are only exceeded by those he makes upon himself. According to my way of looking at things, he involves himself too deeply in the expected outcome of a particular piece of research, so that if in fact it comes out differently, he will be disproportionately worried and spend excessive effort at trying to salvage that which has already proved itself unsalvageable.

Our work on clonus progressed in a manner which was satisfactory to me, but which did not come up to Arturo's rigid requirements for experimental research. The paper has never been published, although I think that many of its ideas have come to be accepted generally.

Besides our study of clonus, we also did some work on the heart as a conductor of rhythmic contractions. This work was later carried to a greater degree of perfection by another one of Arturo's collaborators.

As soon as I could get about freely, I looked up the Vallartas and the mathematicians at the university. Manuel and his wife, Maria Luisa, gave us a warm welcome to Mexico. They lived with a number of brothers, sisters-in-law, nieces, and nephews, in the large house of Maria Luisa's parents, the Margains, on Avenida Insurgentes. Father Margain, who died recently, was a doctor, and his sons were doctors, architects, and lawyers.

Like a number of Mexicans in their circle, my friends Manuel and Maria Luisa were connected with the saga of Maximilian. It had been a great-uncle of Maria Luisa, Lieutenant Margain, who at Querétaro had commanded the squad which shot Maximilian and his two generals Mejía and Montemar. On the other hand, one of her great aunts had been lady-in-waiting to Carlota. An ancestor of Manuel had been governor of a province and a political mainstay of Maximilian's antagonist and conqueror, the Indian president Benito Juárez. Thus, my visit to Mexico made me feel at once as if I had been immersed in the fascinating and violent history of the country.

Another Mexican scientist friend of mine was Nápoles Gándara, who is professor of mathematics at the University of Mexico, and who invited me to lecture there. He had been up to M.I.T. to study with Dirk Struik. His main distinction is not so much the merit of his original research as the way in which he has stood up for many years for mathematics and the training of mathematicians. He is largely and possibly entirely of Indian blood, and he possesses a full share of that modest determination and oneness of purpose which derives from his Indian ancestry.

The Indian element in Mexico and the Spanish differ in many ways. It is the Spaniard rather than the Indian who has the romantic *élan* which we associate with the South. On the other hand, the Indian is unsurpassable where steadiness, loyalty, and conscientiousness come into consideration. Thus, each race furnishes qualities which the country needs. It is a splendid thing that the Indian has come to his own and is participating in the development of a new middle class, which stands on the tripod of the Spaniard, the Indian, and the foreigner.

My Mexican friends came from all of these three elements of society, and it was an exciting thing to see how this new middle class of diverse origins constituted a cordial, friendly, and well-organized body of people.

My friend Erro, the astronomer, took me to see Torres Bodet, the Minister of Education, who was doing much for the education of illiterates. One expects that when, in a Latin country, the head of a national observatory introduces a foreign scholar to the Minister of Education, it will be an affair of frock coats, striped trousers, and great formality. As a matter of fact, Erro appeared in riding breeches and a blue-and-white sweatshirt. Mexico has a charming combination of Spanish and Latin American formality with the free and easy ways of the United States.

I saw a great deal of the medical crowd in Mexico City as well as of the mathematicians, physicists, and astronomers. Everywhere I found a new and active intellectual life in the making. The Mexicans are quite conscious of the distance they must go to come up to the level of the countries older in scientific reputation, but they are determined to make up for their late arrival in scientific history, and the level of work increases year by year. In the meanwhile, there is something peculiarly charming in the friendliness and warmth of heart and in the intellectual devotion of these friends of mine, and Mexico will never seem to me a truly foreign country.

Among my medical friends I wished to speak particularly of Dr. García Ramos, for he in his own personality is a living example of the new Mexico. He was born in Querétaro, of parents in quite modest circumstances, and is overwhelmingly Otomi Indian in blood. He went into the army as a boy. At every stage in which examinations or ability could promote a man's career, he stood out in the first rank. He was sent to the army medical school, from which he was graduated with real distinction. Arturo took him on as his assistant.

García Ramos is now a well-known physiologist in his own right, having received a Guggenheim Fellowship to study in the States. He was a major in the army when I first knew him, and he is now a full colonel. He has retired from active military service, and is at present the head of the Mexico City Nutrition Laboratory. There is no future for a general who does not make the army his first consideration, and García Ramos is infinitely more interested in medical research than in the army. The army has therefore nothing further to offer him.

Among Mexican mathematicians the late Professor G. D. Birkhoff of Harvard has had a great influence. Some years before, he had developed an alternative explanation of certain phenomena which occupy a key position in Einstein's gravitational relativity. Birkhoff's theory, which is not actually relativistic, is meant to account for the displacement of light by the attraction of the sun for certain anomalies in the orbit of Mercury and for the shift of light from the remote corners of the universe toward the red end of the spectrum. At the time at which Birkhoff was working, several of the Mexicans were up in Cambridge in contact with him. They are now passing Birkhoff's influence on to their own students. The new subject became a favorite with the young Mexican school of mathematics, and paper after paper was written on Birkhoff's work. After he died in 1942, the Mexicans continued to carry on the work as a tribute of piety.

I could go on from name to name of my friends without finding a chance to refer to much more than a small part of them,

but there is one person whom I must mention in particular, our
janitor, Olvera, a tall, gaunt peon, who possesses the instinctive
culture which one so frequently finds in strata of Mexican society which tend to be illiterate. By this I mean in the first
instance a pride in a correct and elegant speaking use of the
Spanish language. Indeed, Olvera himself is not illiterate. He
has taken full advantage of his situation as an employee of an
institution of learning to extend his culture in many directions.
In particular, together with several of the younger doctors and
with the stenographers, he has joined an English class conducted by Mrs. Rosenblueth. He has been one of her most
conspicuously successful pupils. Now he is able to speak in an
English which has the same chosen quality that belongs to his
Spanish, and on one occasion he is said to have remarked to
two skylarking American boys in the laboratory, "Gentlemen,
this conduct is not worthy of an international scientist." Indeed, Olvera's choice of words, both in Spanish and in English,
is so notable that when Arturo and I have any matter of
phraseology come up in a paper we are writing, we say, "How
would Olvera have said it?"

Olvera is utterly devoted to the laboratory and to Arturo,
and this devotion may even become embarrassing at times. As
a vigorous man, Arturo naturally prefers to go like anyone else
to the barbershop or the bootblack; but no, Olvera will not let
him. The barber and the bootblack must come to his office at
those times when Olvera has decided that the personal appearance of the boss needs a bit of renovating, and Arturo sits
embarrassedly upon his chair while the door is closed on his
humiliation.

The particular pride of the beautiful modern building of the
Instituto is the pair of mural paintings by Diego Rivera on the
history of the medicine of the heart. For all Rivera's personal
flamboyancy, these paintings show a great depth of serious
study and scholarship. The pieces of medical apparatus which

appear are all correctly delineated. Whether they are of the accepted pattern or not, they all would work. To achieve this, Arturo and Rivera consulted with each other many times.

However, the artistic merit of these paintings goes far beyond such technical details. One of them, which concerns the earlier history of cardiology, has a generally red tone, emanating from the pile of fagots at which Servetus was burned. It must be remembered that, besides being a heretic in the eyes of John Calvin, he was one of the discoverers of the circulation of the blood. Naturally a man like Rivera takes a delight in showing that the burning of heretics was as much a misdeed of the Protestants as of the Catholics.

The other painting is largely blue in coloring, from the light emanating from the Roentgen tube and the other electrical apparatus of the modern cardiologist. In both there is a careful depiction not only of the individuals who have contributed to the science but of their distinct national types. Many single faces or groups are portrayed with great understanding and emotion, and in particular there is a splendid piece which shows the consumptive Laënnec using his stethoscope on a dying heartsick patient. You see the pose of the sick doctor echoing in every line that of the patient, who shows the characteristic Hippocratic face.

The time came for me to go to the mathematical meeting at Guadalajara. I will not say that the mathematics at this meeting was either terrifically novel or very exciting, although it represented a genuine attempt for a country which was rather a newcomer to the mathematical field to do work on a high level. The meeting itself and the city in which it took place were equally charming.

We had quite an American contingent at the meeting, both invited and self-invited. One of the invited guests was Professor Murnaghan, of Johns Hopkins, who found a not unusual difficulty in adjusting himself to the diet of the country. One

morning the Rosenblueths came down early, and Rosenblueth
said in English to nobody in particular, "I feel fine, period!"
I answered, "Murnaghan feels rotten, colon."

We made many excursions about Guadalajara and its surroundings. I was particularly delighted by the sincere and
manly painting of Orozco.

There was a great painting of his in the Governor's Palace,
in which he represented the wars between Fascism and Communism in a powerful but brutal symbolic fashion. However,
the most interesting group of his paintings was in the Hospicio.
This was a public orphanage which seemed to me more human
and far less institutional than most boarding schools. The children were not in uniform but played in ordinary dress with an
abundance of toys in the shady and tree-filled courtyards.
There were two school orchestras, one of older and one of
younger inmates, and the conductor and music teacher, who
had achieved remarkable results on a truly professional level,
was an elderly Indian gentleman with the impassive face that
one has learned to know in the portraits of Juárez.

It was the chapel of this institution that had been decorated
by Orozco, and although the paintings were not the conventional paintings of the Christian tradition, they were most certainly religious paintings, and in the main representations of
a new Apocalypse. They had some of the flat red and blue
colors of El Greco. However, the use of color was not their
strongest point, and the drawing was extremely modern. One
of them showed the wheel crushing the city wall of the Aztecs.
The moral of this was that the Western civilization of the
wheel, which the Aztecs had never known, had crushed the
indigenous cultures. Other panels carried out the theme of the
conquest and showed the Spanish soldiers with their swords
and the monks with their robes. Through the nave of the whole
great church of centuries ago, these paintings carried a spirit
of grim beauty and power, and one felt, for all its grimness

that it was a worthy background against which the children could develop their sense of the nobility of art.

We members of the congress witnessed a dance exhibition by the school teachers of Guadalajara, both men and women. They had nothing schoolmarmish or schoolmasterish about them. The whole show went off with a verve and sincerity which excited our greatest admiration.

When I returned to the States I found that the interest in the sort of work that Arturo and I had been doing together, namely the application of modern mathematical techniques to the study of the nervous system as a problem in communication, had excited a spirited interest. A colleague of mine had persuaded the Macy Foundation, in New York, to organize a number of meetings devoted to this subject. The series ran for several years. Here a group of psychiatrists, sociologists, anthropologists, and the like came together with neurophysiologists, mathematicians, communication experts, and the designers of computing machines, to see if they couldn't find a common basis of thought.

The discussions were interesting and, in fact, we did learn to speak more or less in one another's language, but there were great obstacles in the way of a complete understanding. These semantic difficulties resided in the fact that on the whole there is no other language which can give a substitute for the precision of mathematics, and that a large part of the vocabulary of the social sciences is and must be devoted to the saying of things that we do not yet know how to express in mathematical terms.

Indeed, I found then, as I have found on so many other occasions, that one of the chief duties of the mathematician in acting as an adviser to scientists in less precise fields is to discourage them from expecting too much of mathematics. They must learn that there is no intellectual virtue (and that there is, in fact, a severe intellectual vice) in using a number of three digits when our available accuracy runs to one digit. Thus, while we were quite convinced that the same modes of thought traverse the problems of communication—whether they be so-cial, physiological, or mechanical—it was the mathematicians rather than the physiologists and sociologists who had most to throw cold water on an overestimation of the detailed possi-bilities of mathematics in these other fields.

Arturo attended several of these earlier Macy meetings. We wanted to continue to work together in the intimate way in which we had already started and to secure a backing for this future work that would enable us to keep it up together for a number of years. We managed to interest both M.I.T. and the Instituto Nacional de Cardiología in the project and to secure funds in New York from the Rockefeller Foundation. Here Warren Weaver, who had now returned to his normal work from his war duties, was very enthusiastic and hopeful about the possibilities that had emanated from my research on pre-diction. He represented in this the natural sciences group of the Rockefeller Foundation. Dr. Robert Morison, who repre-sented the biological sciences group, was also interested in the proposition. He was a close friend of Arturo's and had been a member of our dinner group at the Harvard Medical School. Between him and Weaver, we got the signal to go ahead.

M.I.T., the Instituto, and the Rockefeller Foundation came to a decision that I should spend half of every other year in Mexico and that, on the other hand, Arturo should spend part of the intervening years at M.I.T., for a period of five years. With slight modifications we have adhered to this program, and there now remains only one half year of the original pro-gram for Arturo to spend in Boston.

Besides the work Arturo and I did on heart conduction and on clonus, there is a group of biological researches, on some of which he and I have worked together, and some of which have received my independent attention. Most of these have not been carried through to a definitive result, but they still present features of interest for further work.

One piece of work I did with Rosenblueth concerns an attempt to set up and to solve the differential equations of impulse flow along a nerve, and in this manner to compute the passing distribution of electricity which occurs as an impulse goes by. This is the so-called theory of the nerve spike. This sudden rise and sudden drop of potential in the passing of a nerve impulse seemed to me to divide itself into at least three separate consecutive phenomena.

Another research which we undertook together had to do with the statistical theory of the conduction of impulses through a synapse, or a place where incoming nerve fibers join with fibers proceeding further in the nervous system. This was done during one of Dr. Rosenblueth's stays in Cambridge.

Two other researches which have not yet matured but which seem to me to lead in a promising direction have been undertaken by me in collaboration with workers in the electronics laboratory of M.I.T., in particular with Dr. Jerome Wiesner. One of these, in which the leading idea is Wiesner's, concerns an attempt to analyze sounds instrumentally in such a way that the pattern of sounds might be conveyed to the skin as a series of local pressures or vibrations. We made promising headway in the matter but did not come to a definite choice as to just how to pursue the best path in making such apparatus available to the deaf as an alternative mode of registering sound by touch.

This represents one phase of my general interest in the philosophy of prosthesis. I have believed that much could be done with artificial limbs by realizing that the deprivation of the amputee is quite as much sensory as motor, and that the amputee's loss of part of the information which is available to the normal person leaves him to assume a condition which parallels not merely paralysis but ataxia. Ataxia, which is a loss of steering impulses, does not prevent a person from moving, but prevents him from moving in a purposive way, by depriving him of an awareness of his own motions.

Closely related to this concept is the idea of a more adequate iron lung for paralytics. The existing iron lung has saved many lives, but it tends to make the patient dependent on a rigid process of breathing over which he has no control, and it tends to cause him to unlearn the normal process of breathing. There seems to me a real possibility to take off electric signals from such breathing muscles as are not completely dead and to amplify them so that the patient might have the satisfaction of controlling his own iron lung, as well as the exercise of making use of what is left of his breathing muscles. This work awaits the organization of a group of physiologists, physicians, and engineers to carry out the necessary researches.

Of all the things I have done in physiology, that which seems to me most significant is the application to the study of brain waves of the statistical theory of what are known as time series.

My war work on filtering and prediction of time series had represented an extension of my earlier work on generalized harmonic analysis and on the Brownian motion as tools for the study of irregular phenomena distributed in time. For years I had the intention of using these tools in every region in which they seemed apt. From the very beginning of the study of brain waves, ever since I had contact with some of the original electroencephalographers in Arturo's seminar at the Harvard Medical School, I had felt that this field was one in which I could accomplish something; and I have never ceased to importune my neurophysiologist comrades with requests that they give a sympathetic hearing to these methods and that they try them out if possible on some experimental data.

In the early days of brain-wave work, it was supposed that the stray electric currents in the brain, as observed through the scalp, would throw a new light on the physiology of the brain and the associated mental phenomena. Much has in fact been done in the treatment of epilepsy through the study of brain waves, but the great expectations of the thirties have not yet been realized. The reason is that the brain waves as we see them originally are a mixture of very varied phenomena such as we would find if, for example, we observed the stray electric currents around a computing machine or a control machine. They speak a language of their own, but this language is not something that one can observe precisely with the naked eye, by merely looking at the ink records of the electroencephalograph. There is much information contained in these ink records, but it is like the information concerning the Egyptian language which we had in the days before the Rosetta Stone, which gave us the clue to the Egyptian script.

Of recent years, I have been a member of a group involving people from the various laboratories at M.I.T. and from the Massachusetts General Hospital which has been endeavoring to find the Rosetta Stone for the script of the brain waves by means of harmonic analysis. We have had a conspicuous example of success in this field in the past in the work of the great American experimental physicist, Michelson.

Michelson invented an instrument called the interferometer, which was the most delicate machine ever invented for the study of the spectrum of light, and which enabled him to carry out such a seemingly impossible task as the determination of the angles subtended at the earth by some of the fixed stars. The principle of this instrument can in fact be realized in an instrument for the study of brain waves and other such oscillation. We have called this instrument the autocorrelator. Many people at M.I.T., and Lee in particular, have reduced the design of autocorrelators to a surprising degree of perfection.

When the crude original records of brain waves are transformed by the autocorrelator, we obtain a picture of remarkable clarity and significance, quite unlike the illegible confusion of the crude records which have gone into the machine. We are at the very beginning of our work in this field, but we have great hopes of what it will offer for the future; and we should not be surprised if the ambitious expectations of the early electroencephalographers of thirty years ago, of a really legible form of electroencephalograph record, will now begin to be realized.

The analogy between the interferometer and the autocorrelator is deep and significant, and the earlier work of Michelson has given us a whole language for the reading of the results presented by such machines.

I have in fact done the greater part of my share in this brain-wave work in the last three years, since the end of my visit to Mexico, but I consider it essentially as the consummation of the lines of research which Arturo and I had embarked on together.

The autocorrelation study of brain waves is not the only field where my mathematical interests and Rosenblueth's physiological interests have met. The original analogy we found between machine and human feedbacks has been supplemented again and again by striking new analogies which we keep on finding between the nervous system and control or computation machines.

From the very beginning, I was struck by the similarities between the nervous system and the digital computer. I did not mean to claim that these analogies are complete, or that we can exhaust the properties of the nervous system by calling it a digital computer. I merely wish to suggest that certain aspects of its behavior are close to those of the digital computer.

The nervous system is certainly a complicated net of elements which transfer impulses. Fundamentally, if an impulse is strong enough to go from the end of one nerve fiber to the other, it reaches the further end as a whole without much influence of the strength of the impulse at the nearer end, provided that it goes through at all. Thus, the nerve fiber transmits yeses or noes. When an impulse reaches the end of a nerve fiber, it combines with various other impulses that have reached the same level to determine whether the next nerve fiber discharges. In other words, the nerve fiber is a logical machine in which a later decision is made on the basis of the outcome of a number of earlier decisions. This is essentially the mode of operation of an element in a computing machine. Besides this fundamental resemblance, we have auxiliary resemblances pertaining to such phenomena as memory, learning, and the like.

There is one other medical matter that has attracted my attention of recent years. Walter Cannon, going back to Claude Bernard, emphasized that the health and even the very existence of the body depends on what are called homeostatic processes, namely, processes which tend to keep temperature, blood pressure, and the many other factors of the interior environment of a living being so stable that life is possible. That is, the apparent equilibrium of life is an active equilibrium, in which each deviation from the norm brings on a reaction in the opposite direction, which is of the nature of what we call negative feedback.

Thus, when the body goes wrong, there must be cases where the failure is an intrinsic breakdown of the feedback process, and where the mathematical description of the manner of failure indicates the nature of the feedback process and the nature of its breakdown. A colleague of mine, Paul Hahn, and I have applied this sort of discussion to the history of leukemia, and we see considerable evidence that in this excessive growth of the white blood corpuscles there is a homeostatic process balancing the creation and the destruction of the blood cells which is not completely abrogated, but which goes on at an incorrect level. I feel that this concept of a disease of homeostatis may well prove useful in many fields of medicine.

There has been in the past a great tendency in medicine to

think in terms of localization. This has been particularly the case in matters concerning the brain, where a separate function has been discovered or postulated for almost every area of the cortex or surface of the hemispheres. However, the tendency of a strong emphasis on localization has been to subordinate general questions of organization to localizable atomic phenomena.

It seems to me that our studies of control apparatus are giving us a better insight into the way in which these local phenomena are built up into large processes extending all over the brain—or, in fact, the whole human body. In healthy activity these overall processes must be understood, since under pathological conditions they may break down in a manner which cannot be assigned to the failure of the individual parts. There are diseases, like leukemia, where certain processes such as the formation of white corpuscles are apparently running wild. However, even in this diseased activity there are strong signs that what is at fault is not so much an absence of all internal control over the process of corpuscle formation and corpuscle destruction but a control working at a false level.

Most of the pieces of research which Arturo and I undertook during my various visits to Mexico and his to the United States, have already appeared in the technical journals. I shall have more to say of the details of my later visits. My wife and I hope to visit Mexico more times in the future, whether for my research and writing or just to enjoy the life of a country which has been charmingly hospitable to us.

14

Moral Problems
of
a Scientist.
The Atomic Bomb.
1942—

One day during the Second World War I was called down to Washington to see Vannevar Bush. He told me that Harold Urey, of Columbia, wanted to see me in connection with a diffusion problem that had to do with the separation of uranium isotopes. We were already aware that uranium isotopes might play an important part in the transmutation of elements and even in the possible construction of an atomic bomb, for the earlier stages of this work had come before the war and had not been made in the United States.

I went to New York and had a talk with Urey, but I could not find that I had any particular qualification for solving the special problem on which he requested help. I was also very busy with my own work on predictors. I felt that there I had found my niche for the duration of the war. It was a place where my own ideas were particularly useful, and where I did not feel that anyone else could do quite so good a job without my help.

I therefore showed no particular enthusiasm for Urey's problem, although I did not say in so many words that I would not work on it. Perhaps I was not cleared for the problem, or perhaps my lack of enthusiasm itself was considered as a sufficient reason for not using me, but that was the last I heard of the matter. This work was a part of the Manhattan Project and the development of the atomic bomb.

Later on, various young people associated with me were put on the Manhattan Project. They talked to me and to everyone else with a rather disconcerting freedom. At any rate, I gathered that it was their job to solve long chains of differential equations and thereby handle the problem of repeated diffusions. The problem of separating uranium isotopes was reduced to a long chain of diffusions of liquids containing uranium, each stage of which did a minute amount of separation of the two isotopes, ultimately leading in the sum to a fairly complete separation. Such repeated diffusions were necessary to separate two substances as similar in their physical and chemical properties as the uranium isotopes. I then had a suspicion (which I still have, though I know nothing of the detail of the work) that the greater part of this computation was an expensive waste of money. It was explained to me that the effects on which one was working were so vanishingly small that without the greatest possible precision in computation they might have been missed altogether.

This however did not look reasonable to me, because it is exactly under these circumstances of the cumulative use of processes which accomplish very little each time that the standard approximation to a system of differential equations by a single partial differential equation works best. In other words, I had and have the greatest doubts, to the effect that in this

very slow and often repeated diffusion process those phenomena which may not be justifiably treated as continuous are
of very slight real importance.

Be that as it may, while I did not have any detailed knowledge of what was being done on Manhattan Project, the time
came when neither I nor any other active scientist in America
could fail to be aware that such a project was under way. Even
then we did not have any clear idea of how it was to be used.
We were afraid that the main use to be made of radioactive
isotopes was as poisons. We feared that here we might well find
ourselves in the position of having developed a weapon which
international morality and policy would not permit us to use,
even as they had held the Germans back from the use of poison
gas against cities. Even were the work to result in an explosive,
we were not at all clear as to the possibilities of the bomb nor
as to the moral problems which its use would involve. I was
very certain of at least one thing: that I was most happy to
have had no share in the responsibility for its development
and its later use.

So far the moral problem of warfare had not concerned me
directly. However, in the fall of 1944 a complex of events took
place which had a very considerable effect on my later career
and thought. I had already begun to reflect on the relation
between the high-speed computing machine and the automatic
factory, and I had come to the conclusion that as the essence
of the computing machine lay in its speed and in its programming, or determination of the sequence of operations to be
performed by means of a magnetic tape or punched cards, the
automatic factory was not far off. I wondered whether I had
not got into a moral situation in which my first duty might be
to speak to others concerning material which could be socially
harmful.

The automatic factory could not fail to raise new social
problems concerning employment, and I was not sure that I

had the answers. A vast redistribution of labor at different levels would be created. When the human being is being used mechanically, simply as an inferior sort of switching or decision device, the automatic factory threatens to replace him completely by mechanical agencies. On the other hand, it creates a new demand for the highly skillful professional man who can organize the order of operations which will best serve a particular function.

It will also create a demand for trouble-shooters and maintenance crews of a particularly well-trained sort. If these changes in the demand for labor come upon us in a haphazard and ill-organized way, we may well be in for the greatest period of unemployment we have yet seen. It seemed to me then quite possible that we could avoid a catastrophe of this sort, but if so, it would only be by much thinking, and not by waiting supinely until the catastrophe is upon us. I shall say something later in this chapter of my present opinion in the matter.

Accordingly, when a colleague wished for some information concerning my newer work I answered that I was not by any means certain that this work should be communicated to him, or indeed to the public at large. I felt all the more strongly about this inasmuch as he had requested the information for military purposes, and I did not know whether I should be a party to the use of my new ideas for controlled missiles and the like.

I showed the letter to a colleague of mine who happened to have a flair for journalism. He immediately suggested that I send my note to the *Atlantic Monthly* as a basis for what might be a more elaborate article. I followed his suggestion and sent it. If I had thought out fully how I was thus subjecting myself to a deep moral committment while he was subjecting himself to nothing at all, I might well have hesitated, although I probably would put this hesitation behind me as an act of cowardice. The moral consequences of my act were soon to follow.

About this time I had agreed to participate in two meetings: the earlier one on applied mathematics, called together by Princeton at the end of its second hundred years of existence; and a later one organized by Aiken at Harvard on the subject of automatic high-speed computing machines. To this second meeting, which took place under the joint auspices of Harvard University and the Navy Department, I had agreed to give a paper. Meanwhile I reported at the Princeton meeting on my work on prediction theory. While my talk covered material which I knew could ultimately be used for military purposes, I had counted on the abstractness of my presentation and the natural inertia of many of my colleagues to prevent the work from being put to immediate and uncontrollable military use.

However, my hand was forced. A colleague who had previously taught at M.I.T. and had gone back to his home University of California, because of his earlier associations there and because of the climate, had been pushing my name as head of or consultant for a military or semimilitary project on mechanical computation, to be located in California. He had not consulted me in the matter, but he had assumed that an invitation to California would be accepted by me without question.

I have said that the project was semimilitary. It was in fact to be under the Bureau of Standards, but it was quite clear that all the facilities engineered by the project if it should be successful would be pre-empted for years by the military services. My acquaintance had tried to commit me not only to work whose objective was distasteful to me but to work which would involve conditions of secrecy, of a police examination of my opinions, and of the confinements of administrative responsibility. These I could not accept. They would have bound me to a course of conduct which would have broken me down in a very few months. When the invitation was passed on to me, I considered my *Atlantic Monthly* letter, and I had no alternative but to say no. I probably would have said no anyway, as I did not fancy myself as an administrator.

Then I recalled the military meeting at Harvard at which I had already committed myself to speak. There were some two weeks before it would come off, so I thought that I had ample time to change my policy. I went to Aiken and tried to explain the situation. In particular, I pointed out to him that the California offer had made it necessary for me to take a definite stand on my war work and that I could not accept one sort of a military association and reject another. I therefore asked to be released from my promise to give a talk.

I gathered from Aiken that there would be time to take my name off the list of speakers for the meeting. However, when the meeting came, I found that Aiken had done this by merely running a line through my name on the printed programs which had been issued to members of the meeting and to the press.

This procedure was extremely embarrassing to me. It was even more embarrassing to him. The newspapermen came to me and asked whether this striking out of my name had anything to do with the letter that had appeared in the *Atlantic Monthly*. I said that it had, and I tried to explain to them the circumstances that had forced my hand. I took full responsibility in the matter and said that I acted with Aiken's consent and that I was not taking this step out of pique or personal animosity.

Naturally, they went to Aiken in the matter. Without reflection, he assumed that I had been involved in some deep plot to discredit him and to turn the meeting into a public scandal. In fact, as I have said, I had consulted him at the very beginning and had understood that there was to be no publicity about the affair. The matter would have had no publicity if it had not been for the emphasis which he had placed on my participation in the meeting by the way he scratched out my name.

All these emotional experiences were nothing to those through which I went at the time of the bombing of Hiroshima. At first I was of course startled, but not surprised, as I had been aware of the possibility of the use of the new Manhattan Project weapons against an enemy. Frankly, however, I had been clinging to the hope that at the last minute something in the atomic bomb would fail to work, for I had already reflected considerably on the significance of the bomb and on the meaning to society of being compelled to live from that time on under the shadow of the threat of limitless destruction.

Of course I was gratified when the Japanese war ended without the heavy casualties on our part that a frontal attack on the mainland would have involved. Yet even this gratifying news left me in a state of profound disquiet. I knew very well the tendency (which is not confined to America, though it is extremely strong here) to regard a war in the light of a glorified football game, at which at some period the final score is in, and which we have to count as either a definite victory or a definite defeat. I knew that this attitude of dividing history into separate blocks, each contained within itself, is by no means weakest in the Army and Navy.

But to me this episodic view of history seemed completely superficial. The most important thing about the atomic bomb was, in my opinion, not the termination of a specific war without undue casualties on our part, but the fact that we were now confronted with a new world and new possibilities with which we should have to live ever after. To me the most important fact about the wars of the past was that, serious as they had been, and completely destructive for those involved in them, they had been more or less local affairs. One country and one civilization might go under, but the malignant process of destruction had so far been localized, and new races and peoples might take up the torch which the others had put down.

I did not in the least underrate the will to destructiveness, which was as much a part of war with a flint ax and of war with a bow and arrow as it is of war with a musket and of war with a machine gun. What came most strongly to my attention was that in previous wars the power of destruction was not commensurate with the will for destruction. Thus, while I realized that as far as the people killed or wounded are concerned, there is very little difference between a cannonade or an aerial bombardment with explosive bombs of the type already familiar and the use of the atomic bomb, there seemed to me to be most important practical differences in the consequences to humanity at large.

Up to now no great war, and this includes World War II, had been possible except by the concerted and prolonged will of the people fighting, and consequently no such war could be undertaken without a profoundly real share in it by millions of people. Now the new modes of mass destruction, expensive as they must be in the bulk, have become so inexpensive per person killed that they no longer take up an overwhelming share of a national budget.

For the first time in history, it has become possible for a limited group of a few thousand people to threaten the absolute destruction of millions, and this without any highly specific immediate risk to themselves.

War had made the transition between an overwhelming assertion of national effort and the push-button declaration of the will of a small minority of people. Fundamentally this is true, even if one includes in the military effort all the absolutely vast but relatively small sums which have been put into the whole body of nuclear research. It is even more devastatingly true if one considers the relatively minimal effort required on the part of a few generals and a few aviators to place on a target an atomic bomb already made.

Thus, war has been transported, at least as a possibility, from the field of national effort to the field of private conspiracy. In view of the fact that the great struggle to come threatens to be one between the United States and the Soviet Government, and in view of the additional fact that the whole atmosphere and administration of the Soviet Government shares with that of the Nazis an extremely strong conspiratorial nature, we have taken a step which is intrinsically most dangerous for us.

I did not regard with much seriousness the assertions which some of the great administrators of science were making, to the effect that the know-how needed for the construction of the atomic bomb was a purely American thing and could not be duplicated by a possible enemy for many years at least, during which we could be counted upon to develop a new and even more devastating know-how. In the first place, I was acquainted with more than one of these popes and cardinals of applied science, and I knew very well how they underrated aliens of all sorts, in particular those not of the European race. With my wide acquaintance among scholars of many races and many countries, I had not been able to discern that scientific ability and moral discipline were the peculiar property of those of blanched skin and English speech.

But this was not all. The moment that we had declared both our possession of the bomb and its efficiency by using it against an enemy, we had served notice on every country that its continued existence and independence of policy were conditioned on its prompt possession of a similar weapon. This meant two things: that any country which was our rival or potential rival was bound to push nuclear research for the sake of its own continued independent existence, with the greatly stimulating knowledge that this research was not intrinsically in vain; and that any such country would inevitably set up an espionage system to get hold of our secrets.

This is not to say that we Americans would not be bound in self-defense to oppose such leaks and such espionage with our full effort for the sake of our very national existence; but it does mean that such considerations of legality and such demands on the moral responsibility of loyal American citizens could not be expected to have the least force beyond our frontiers. If the roles of Russia and the United States had been reversed, we should have been compelled to do exactly what they did in attempting to discover and develop such a vital secret of the other side; and we should regard as a national hero any person attached to our interests who performed an act of espionage exactly like that of Fuchs or the Rosenbergs.

I then began to evolve in my mind the general problem of secrets; not so much as a moral issue, but as a practical issue and a policy which we might hope to maintain effectively in the long run. Here I could not help considering how soldiers themselves regard secrets in the field. It is well recognized that every cipher can be broken if there is sufficient inducement to do it and if it is worthwhile to work long enough; and an army in the field has one-hour ciphers, twenty-four-hour ciphers, one-week ciphers, perhaps one-year ciphers, but never ciphers which are expected to last an eternity.

Under the ordinary circumstances of life, we have not been accustomed to think in terms of espionage, cheating, and the like. In particular, such ideas are foreign to the nature of the true scientist, who, as Einstein has pointed out, has as his antagonist a world which is hard to understand and interpret, but which does not maliciously and malignantly resist this interpretation. "The Lord is subtle, but he isn't plain mean."

With ordinary secrets of limited value, we do not have to live under a perpetual fear that somebody is trying to break them. If, however, we establish a secret of the supreme value and danger of the atomic bomb, it is not realistic to suppose that it will never be broken, nor that the general good will among scientists will exclude the existence of one or two who, either because of their opinions or their slight resistance to moral pressure, may give our secrets over to those who will endanger us.

If we are to play with the edged tools of modern warfare, we are running not merely the danger of being cut by accident and carelessness but the practical certainty that other people will follow where we have already gone and that we shall be exposed to the same perils to which we have exposed others. Secrecy is thus at once very necessary and, in the long run, quite impossible. It is unrealistic to give over our main protection to such a fragile defense.

There were other reasons, moreover, which, on much more specific grounds, made me feel skeptical of the wisdom of the course we had been pursuing. It is true that the atomic bomb had been perfected only after Germany had been eliminated from the war, and that Japan was the only possible proving ground for the bomb as an actual deadly weapon. Nevertheless, there were many both in Japan and elsewhere in the Orient who would think that we had been willing to use a weapon of this terribleness against Japan when we might not have been willing to use it against a white enemy. I myself could not help wondering whether there might not be a certain degree of truth in this charge. In a world in which European colonialism in the Orient was rapidly coming to an end, and in which every oriental country had much reason to be aware of the moral difference which certain elements in the West were in the habit of making between white people and colored people, this weapon was pure dynamite (an obsolete metaphor now that the atomic bomb is here) as far as our future diplomatic policy was concerned. What made the situation ten times worse was that this was the sort of dynamite which Russia, our greatest potential antagonist if not our greatest actual enemy, was in a position to use, and would have no hesitation whatever in using.

It is the plainest history that our atomic bomb effort was international in the last degree and was made possible by a group of people who could not have been got together had it not been for the fact that the threat of Nazi Germany was so strongly felt over the world, and particularly by that very scholarly group who contributed the most to Nuclear theory. I refer to such men as Einstein, Szilard, Fermi, and Niels Bohr. To expect in the future that a similar group could be got together from all the corners of the world to defend our national policy involved the continued expectation that we should always have the same moral prestige. It was therefore doubly unfortunate that we should have used the bomb on an occasion on which it might have been thought that we would not have used it against white men.

There was another matter which aroused grave suspicion in the minds of many of us. While the nuclear program did not itself involve any overwhelming part of the national military effort, it was still in and for itself an extremely expensive business. The people in charge of it had in their hands the expenditure of billions of dollars, and sooner or later, after the war, a day of reckoning was bound to come, when Congress would ask for a strict accounting and for a justification of these enormous expenditures. Under these circumstances, the position of the high administrators of nuclear research would be much stronger if they could make a legitimate or plausible claim that this research had served a major purpose in terminating the war. On the other hand, if they had come back empty-handed—with the bomb still on the docket for future wars, or even with the purely symbolic use of the bomb to declare to the Japanese our willingness to use it in actual fact if the

war were to go on—their position would have been much
weaker, and they would have been in serious danger of being
broken by a new administration coming into power on the
rebound after the war and desirous of showing up the graft
and ineptitude of its predecessor.

Thus, the pressure to use the bomb, with its full killing
power, was not merely great from a patriotic point of view
but was quite as great from the point of view of the personal
fortunes of people involved in its development. This pressure
might have been unavoidable, but the possibility of this pressure, and of our being forced by personal interests into a
policy that might not be to our best interest, should have been
considered more seriously from the very beginning.

Of the splendid technical work that was done in the construction of the bomb there can be no question. Frankly, I can
see no evidence of a similar high quality of work in the policy-
making which should have accompanied this. The period between the experimental explosion at Los Alamos and the use
of the bomb in deadly earnest was so short as to preclude the
possibility of clear thinking. The qualms of the scientists who
knew the most about what the bomb could do, and who had
the clearest basis to estimate the possibilities of future bombs,
were utterly ignored, and the suggestion to invite Japanese
authorities to an experimental exhibition of the bomb somewhere in the South Pacific was flatly rejected.

Behind all this I sensed the desires of the gadgeteer to see
the wheels go round. Moreover, the whole idea of push-button
warfare has an enormous temptation for those who are confident of their power of invention and have a deep distrust of
human beings. I have seen such people and have a very good
idea of what makes them tick. It is unfortunate in more than
one way that the war and the subsequent uneasy peace have
brought them to the front.

All these and yet other ideas passed through my mind on

the very day of Hiroshima. One of the strong points and at
the same time one of the burdens of the creative scholar is
that he must stand alone. I wished—oh how I wished!—that I
could be in a position to take what was happening passively,
with a sincere acceptance of the wisdom of the policy makers
and with an abdication of all personal judgment. The fact
is, however, that I had no reason to believe that the judgment of these men on the larger issues of the situation was
superior to my own, whatever their technical information
might be. I knew that more than one of the high officials of
science had not one tenth my contact with the scientists of
other countries and of other standpoints and was in nowhere
nearly as good a position to assess the world reaction to the
bomb. I knew, moreover, that I had been in the habit of
considering the history of science and of invention from a
more or less philosophic point of view, and I did not believe
that those who made the decisions could do this any better
than I might. The sincere scientist must back his bets and
guesses, even when he is a Cassandra and no one believes him.
I had behind me many years of lonely work in science where
I had finally proved to be in the right. This inability to trust the
Powers That Be was a source of no particular satisfaction to
me, but there it was, and it had to be faced.

One of my greatest worries was the reaction of the bomb
on science and on the public's attitude to the scientist. We
had voluntarily accepted a measure of secrecy and had given
up much of our liberty of action for the sake of the war, even
though—for that very purpose, as many of us thought—more
secrecy than the optimum was imposed, and this at times had
hampered our internal communications more than the informa-
tion-gathering service of the enemy. We had hoped that this
unfamiliar self-discipline would be a temporary thing, and
we had expected that after this war—as, after all, before—we
should return to the free spirit of communication, intranational

and international, which is the very life of science. Now we found that, whether we wished it or not, we were to be the custodians of secrets on which the whole national life might depend. At no time in the forseeable future could we again do our research as free men. Those who had gained rank and power over us during the war were most loath to relinquish any part of the prestige they had obtained. Since many of us possessed secrets which could be captured by the enemy and could be used to our national disadvantage, we were obviously doomed to live in an atmosphere of suspicion forever after, and the police scrutiny on our political opinions which began in the war showed no signs of future remission.

The public liked the atomic bomb as little as we did, and there were many who were quick to see the signs of future danger and to develop a profound consciousness of guilt. Such a consciousness looks for a scapegoat. Who could constitute a better scapegoat than the scientists themselves? They had unquestionably developed the potentialities which had led to the bomb. The man in the street, who knew little of scientists and found them a strange and self-contained race, was quick to accuse them of a desire for the power of destruction manifested by the bomb. What made this both more plausible and more dangerous was the fact that, while the working scientists felt very little personal power and had very little desire for it, there was a group of administrative gadget workers who were quite sensible of the fact that they now had a new ace in the hole in the struggle for power.

At any rate, it was perfectly clear to me at the very beginning that we scientists were from now on to be faced by an ambivalent attitude. For the public, who regarded us as medicine men and magicians, was likely to consider us an acceptable sacrifice to the gods as other, more primitive publics do. In that very day of the atomic bomb the whole pattern of the witch hunt of the last eight years became clear, and what we are living through is nothing but the transfer into action of what was then written in the heavens.

While I had no share in the atomic bomb itself, I was nevertheless led into a very deep searching of soul. I have already explained how my work on prediction and on computing machines had led me to the basis of cybernetics, as I was later to call it, and to an understanding of the possibilities of the automatic factory. From the strictly scientific point of view, this was not as revolutionary as the atomic bomb, but its social possibilities for good and for evil were enormous. I tried to see where my duties led me, and if by any chance I ought to exercise a right of personal secrecy parallel to the right of governmental secrecy assumed in high quarters, suppressing my ideas and the work I had done.

After toying with the notion for some time, I came to the conclusion that this was impossible, for the ideas which I possessed belonged to the times rather than to myself. If I had been able to suppress every word of what I had done, they were bound to reappear in the work of other people, very possibly in a form in which the philosophic significance and the social dangers would be stressed less. I could not get off the back of this bronco, so there was nothing for me to do but to ride it.

I thus decided that I would have to turn from a position of the greatest secrecy to a position of the greatest publicity, and bring to the attention of all the possibilities and dangers of the new developments. I first thought of the trade unions as the people who would naturally be most interested in the matter. My friends directed me towards two union leaders, one of them an intellectual counselor who had himself very little direct authority among the union people with whom he was associated, and the other a high official of the typographers' union. In both cases I found a confirmation of what my English friends had told me some years before: The union official comes too directly from the workbench, and is too immediately concerned with the difficult and highly technical problems of shop stewardship, to be able to entertain any very forward-looking considerations of the future of his own craft.

I found plenty of good will among my union friends but an absolute block on their part to communicate my ideas to their union workers. This was in the middle forties; since then the situation has changed radically. I have been in repeated communication with Mr. Walter Reuther, of the United Automobile Workers, and I have found in him both an understanding of my problems and a willingness to give my ideas publicity through his union journals. In fact, I have found in Mr. Reuther and the men about him exactly that more universal union statesmanship which I had missed in my first sporadic attempts to make union contacts.

There is another quarter in which the sort of ideas I have had concerning the automatic factory have made gratifying headway. This is in the circles of management itself. In the winter of 1949 I gave a talk to the Society for the Advancement of Management concerning the automatic factory as a technical possibility and the social problems it would introduce, and in both matters I was backed up by high management authorities, as, for example, an executive of Remington Rand, Inc. In December of 1952 I was asked to give a talk on a similar subject as part of a symposium on the automatic factory held by the American Society of Mechanical Engineers.

The progress in the general attitude from the first talk to the second was remarkable. Not only was the attending public much larger and my technical remarks confirmed by automatic-machine men for several industries, but the social consciousness of the group as a whole was far beyond what I had found three years before.

While there were a good many who were more sanguine than I had been as to the possibility of achieving a large measure of industrial automatization without catastrophe, there was a general awareness of the interest of the public at large in a meeting which was going to affect so profoundly their future method of life. In particular, problems of the grade-up of repetitive factory workers into trouble-shooting men (and indeed into a sort of junior engineer) occupied a great deal of attention.

Another much-debated problem was that of the new leisure we might expect in the future and the use that could and must be made of it. Indeed, I heard hard-boiled engineering administrators express views which sounded remarkably like the writings of William Morris. Above all, I had everyone backing me in cautioning that the new displacement of human beings from the repetitive labor of the factory must not be taken as a devaluation of the human being and a glorification of the gadget.

The years that have passed since this talk have seen the automatic factory develop from a remote possibility into a beginning actuality, and we can start to assess on a factual basis its probable impact on society. The first industrial revolution of the early nineteenth century replaced the individual by the machine as a source of power. No factory worker of the present day is earning any large part of his wages by the horsepower of his output. Even if he is doing the hardest sort of physical labor, as for example, if he is a steel puddler, his pay is not primarily given him as a prime mover in a power process. What he is actually paid for is his experience and knowledge of how to exert his strength most effectively in a highly purposeful manufacturing process.

However, the strong men of industry such as the steel puddlers are in a decided minority. The factory worker finds a small electric motor or a pneumatic tool at his elbow, and these will give him the sheer physical strength of ten men.

His business is to accomplish a certain purpose by going through certain motions in a given succession. If, for example, he is pasting labels on tin cans, he must see that he has the correct stack of labels before him, that he has moistened them correctly, that he has put them in the correct position on the can, and that he turns at the proper time from one can to the next. This sort of laborer goes through a purely repetitive process, making the minimum demands on any but the lowest level of judgment and observation.

Of course, there are other forms of factory labor. There are the foremen and there are the members of the trouble-shooting gangs, who at the very lowest level must be skilled craftsmen and, on the higher levels, are assimilated in their function to junior engineers. Leaving out these higher ranks of labor, the routine factory worker is often doing so conventional a task that every motion of his and the cue for every motion, may be assigned in advance. This is the point of such efficiency systems as the motion study of Taylor and the Gilbreths.

I have already indicated that it is this level of work which will be replaced by the operations of the automatic factory. Essentially, to my way of seeing things, most of the human labor which the automatic factory displaces is an inhuman sort of human labor, which has been considered a natural task for human beings only since the historical accident of the industrial revolution. Nevertheless, any sudden and uncompensated displacement of this labor must have catastrophic consequences in the direction of unemployment.

Where will this labor have to go? The most obvious answer is that even the automatic factory will always require a considerable group of trouble shooters, skilled craftsmen, and specialists in programming or in the adaptation of the machines to specific problems. During a gradual process of automatization, the natural place for unskilled factory labor to go is into these higher cadres, by some sort of up-grading. The question of the possibility of this up-grading thus becomes vital.

There is a considerable amount of evidence that the sources of labor which furnished the unskilled factory labor of the past generation are drying up, because, since soon after the end of the First World War, we have had no extensive body of immigrants seeking to establish and settle themselves in the country and willing to accept any degree of economic under-valuing. It is the children of this last extensive generation of immigrants who fought in the Second World War, and the rising generation of the present day consists of their children's children. These younger generations are unwilling to accept the permanent position of economic inferiority belonging to the unskilled workers in the old type of factory. Many of them are going into the professions, and even those who are not are beginning to demand that their work be interesting and not a blind alley.

This is not the first time in our industrial history in which technical advances have been conditioned by the decreasing availability of labor of a certain type. Automatic telephone switching came in simply because the old system of hand switching bade fair to demand the entire population of girl high school graduates.

Another matter which may make the stepping up of labor easier than it might have seemed a few years ago is the train-ing of a very considerable part of the young men in our military services as technicians of a relatively high grade. This has been particularly the case in the Air Force. The sort of young man who can be trained to direct and to care for a radar instrument is certainly the sort who can easily learn to be a member of a factory trouble-shooting gang.

Thus it is quite possible, although it is not certain, that the labor environment for the automatic factory has come just at the right time. At any rate, the atmosphere into which the automatic factory is coming is one where it fits into a definite niche of human activity, and one which has been alerted to both the advantages of automatization and the risks.

While this has been the work of many hands, I feel proud that some part of the healthy and understanding atmosphere into which automatization is coming, and of the collaboration of labor and management in being prepared to work out jointly a mode of industrial life which embraces the automatic factory, may be due to my early efforts to alert both of these elements.

15

Nancy,

Cybernetics,

Paris, and After:

1946—1952

In the summer of 1946 there was to take place a private mathematical conference on harmonic analysis in France, at the University of Nancy. I was invited to participate. As a matter of fact, much of the meeting was to deal with my ideas. I traveled to England on a Dutch boat, and before taking part in the meeting I made my usual visit to England and my English friends. I visited University College, London, where J. B. S. Haldane was teaching. He had been divorced from his first wife, and he was now married to a brilliant young geneticist who had been his assistant during the war in physiological experiments they had made concerning the effect on various gases under high pressure.

Both of them had repeatedly put on diving suits and descended into steel tanks of water, and had been subjected to concentrated atmospheres of different gases until the gases had become so poisonous that they had gone into convulsions. I believe Haldane had gone into convulsions four times and his wife seven times. This was in keeping with the tradition which Haldane had followed of subjecting himself as his own guinea pig to every extreme physiological condition which he could find; and of his intrepidity earlier in the war, which made him specialize in the opening and disarming of enemy mines cast ashore.

In general, Haldane is the type of person who deliberately puts himself into positions of danger, discomfort, or unpopularity when there is a job to be done which he considers important. There is something in him, on a more rational level, similar to the reckless drive I had also found in my colleague Paley.

While staying with the Haldanes, I had a good time visiting my colleagues at the National Physical Laboratories at Teddington, the various colleges of the University of London, the University of Manchester, and at Cambridge. I found that Manchester was well at the front of the new technique of high-speed automatic computing machines. At the National Physical Laboratories, Turing was making the same sort of synthesis between mathematical logic and electronics which Shannon had made in the United States. In short, I found the British atmosphere entirely ripe for the assimilation of the new ideas which I was then developing concerning control, communication, and organization.

Actually, the idea of a comprehensive book on these subjects began to come to a head when I reached Paris. There an M.I.T. colleague introduced me to one of the most interesting men I have ever met, the publisher Freymann, of the firm of Hermann et Cie.

Freymann, who, alas, has just died, was a Mexican, and he first came to Paris as a cultural attaché in the Mexican diplomatic service. One of his grandfathers was a retired German sea captain, who had made his home in the region of Tepic, on the west coast of Mexico. The other grandfather was a Huichol Indian chieftain from the same region. My friend's two grandmothers were both of Spanish origin. Freymann, who kept a drab little bookshop opposite the Sorbonne, where every now and then one of the scientific or other intellectual notables would drop in, delighted in relating to me the way each of his two grandfathers tried to capture him from the influence of the other, the one telling him always to be European, and the other reminding him that he was an Indian.

We talked at length about Mexico, and finally the talk came around to my scientific work. Freymann broached the matter which had chiefly interested him. Would I write up my ideas concerning communication, the automatic factory, and the nervous system as a booklet for his series?

He explained to me that he was the son-in-law of the former publisher Hermann, and that he had been the only one of the family to wish to continue the firm when his father-in-law had died. He told me of the various devices by which he had secured the publication contracts for a number of societies, and how he had used this to build up a really intellectual publishing house, as nearly free from the motive of profit as any publishing house can be.

I had already heard of the singular group of French mathematicians who pooled their efforts with the pseudonym "*Bourbaki*," the result of a student hoax, when the group had started to write under the name of a French general of former days. Freymann told me that he was in fact the founder of the group and that he wished to extend it by furthering a new fictitious university, called the University of Nancago, after the two existing schools at Chicago and Nancy.

I felt that it would be fun to get in with so interesting a

group. I agreed to do a book for Freymann, and we sealed the contract over a cup of cocoa in a neighboring *patisserie*.

During all this time I was in contact with Mandelbrojt, who in fact was the organizer of the Nancy congress. We did some mathematical work together, and he accompanied me to Nancy in the little high-speed rail car that now takes the place of an express between the capital and that city. I was put up with the other foreign bigwigs at the beautiful hotel which forms a part of the harmonious square of buildings at the Place Stanislaus.

The glory of the square dates back to the eighteenth century, when the ex-King of Poland became the Duke of Lorraine. Nancy almost rivaled Paris or Versailles as a capital. They say that the etiquette of the court of Nancy was even stricter than that of Versailles. In fact, this ultimately led to the death of the duke. One day, it is said, while walking on the roof and a bit tipsy, he fell down one of the chimneys, and as no person of sufficient rank was available to touch the royal person, he had to stay there until he suffocated.

The hotel at which I stayed was the headquarters for the foreign visitors. There was Harald Bohr, from Denmark; Carleman, from Sweden; Ostrowski, from Basel, and dear old Papa Plancherel from the Zurich Federal Institute of Technology. Jessen was there from Denmark and Beurling from Sweden, both of whom belonged to a younger generation.

Of these, Harald Bohr and Carleman are now dead. Carleman's death was peculiarly tragic, as it so typically followed a Scandinavian pattern which is familiar to those who know the plays of Ibsen and Strindberg. He died of drink—not the social drinking which leads so often to ruin here—but the fiery, passionate dipsomania which is a common disease even in the very best circles of the Scandinavian countries. During the meeting he was often a bit drunk, and afterwards in Paris

I saw him come to Mandelbrojt's apartment for an advance on the travel money due him, red-eyed, with a three-day beard.

Of all the Nancy people, the one whom I saw most often was Laurent Schwartz. His wife was Paul Lévy's daughter, whom I had met years before when I visited her father at Pougues-les-Eaux. Schwartz was active along lines very similar to my own. He had generalized still further the field which I had already treated in my *Acta* paper on generalized harmonic analysis. He reduced it to that highly abstract basis which is characteristic of all the work of the Bourbaki school to which he belonged.

As individuals and as a group, we guests were taken into the small-city social life of Nancy. It was a difficult and austere time in France, when wine was replaced by grape juice, and the delicious French bread loaded with perhaps fifty per cent of maize flour; but our hosts outdid one another in their hospitality. It was quite clear that if one hostess had three sorts of cakes at her party, the next would be constrained to have four, and the next five. At these parties we met M. le Recteur, M. le Maire, M. le Préfet. We were left with the conviction that M. le Recteur, M. le Maire, M. le Préfet, together with their respective ladies, had been meeting one another week in and week out for years. With all the courtesy, culture, and education that this life manifests, its stiffness would make a small New England town look like a sanctuary of social freedom.

At the time of which I write, the University of Nancy had been less damaged by the centralizing pull of Paris than many others. Schwartz has since gone to the capital, after the conventional pattern of French academic careers. At the time of my visit, however, Nancy was an excellent center for young visiting mathematicians who wished to see French university life at its best and who wished to get the full attention of young men still at their greatest vigor and on the way up. Now it again shows signs of relapsing into some of the French provincial apathy.

The meeting was a highly successful one, and we found it possible to integrate our work very well. I returned to Paris. After a few days' work with Bouligand, in which we started a joint paper, and further consultations with Freymann, I crossed over to England and did a little brief visiting before I took the boat from Southampton. I think it was on this second trip that I visited Oxford again and that I went even further west to Bristol, where I saw Grey Walter, who showed me the exceedingly interesting work he was doing on electroencephalography.

Richard Caton had made electroencephalographic studies on animals in England in 1875, but a German, Hans Berger, was the first to make observations concerning certain electrical potentials which displayed themselves on the human scalp. These potentials have their origin in the electrochemical activity of the brain and show a certain relation to various neural and mental disorders which is not too clearly decipherable. Originally, there were great expectations of these phenomena as a direct mode of access to the physiology of the brain, and they have in fact shown certain readable characteristics in cases of epilepsy and of threatening epilepsy.

Apart from this, there are certain regularities of brain waves which can be observed under the proper conditions. The most conspicuous and consistent of these is the so-called alpha rhythm, which is an oscillation of a period of about a tenth of a second.

The art of reading such irregular oscillations is not easy, and there is much that they contain which is not accessible to the naked eye. As I have said in an earlier chapter concerning my experiences on work in physiology jointly with Arturo Rosenblueth, I have recently developed mathematical tools which enable the observer to make more nearly definite statements concerning brain waves. This work is now under way as a joint undertaking in which scientists from M.I.T. and the Massachusetts General Hospital collaborate.

Dr. Grey Walter, although an American by origin, has long lived in Europe, and may be considered one of the leaders of the European study of brain waves or electroencephalography. He is a man who is full of enthusiasm and energy, and he has devised apparatus to give a comprehensive picture of brain waves in different parts of the brain. This picture is interesting and will without any doubt be useful in the study of the normal physiology of the brain and the diagnosis of its disorders. It is, however, more synoptic and less precise in mathematical detail than the tool which we are using in our researches. In fact, Walter's view of science is closer to that of a graphic artist than to that of a mathematician.

At about the same time as I myself, Walter had begun to see the analogies between the feedback machine and the human nervous system and to construct mechanisms which would exhibit some of the features of animal behavior. I have been working on the "moth" which would automatically steer itself into a light. Walter's automata were called "turtles," and had a more complicated repertory of behavior. This included a mechanism by which they might avoid one another in their motions and another mechanism by which when they were "hungry"—literally, when their storage batteries had been ex-hausted—they might be directed into something analogous to a rabbit hutch, where they could feed themselves on electricity until their storage battery was recharged.

I returned home from Southampton on a Dutch ship, the same one I had taken on the way over. On both voyages a large part of the passenger list were Dutch peasants who had immi-grated to the United States, many of whom had lived in the Michigan region around Grand Rapids. They were mostly farmers by origin, and of that intensely religious Calvinistic upbringing which one sees so much in Holland. They had been home after the war for a first contact with their kinsmen who had suffered greatly in the struggle, and they played an important part in the restoration of the country. I am afraid that I forfeited much of their esteem by swearing once or twice in Dutch, and the shiver that passed over them was of a quality that I would not have seen even in the New England countryside.

Nevertheless, when these fine, simple, dignified, respectable people got into the smoking room and had a glass or two of Holland gin under their belts, they would sing the old Dutch songs and even start to dance in the old Dutch manner which is shown in the paintings of Jan Steen, Adriaen Brouwer, and the elder Brueghel. The clothes had changed, but the faces of these staid farm women and vigorous, tight-faced farmers had not changed at all, and even some of the songs, and I suppose some of the dances, dated back to the seventeenth century.

When I got back to the States I found that I had to take up again my work in Mexico. My daughter Barbara was a bit undecided what to do that summer, and I took her down with me. I started more neurophysiological work with Arturo Rosenblueth, and I continued to live among the same group and in very much the same manner as on my previous visits. Barbara and I, and later Margaret, who joined us, found quarters in an apartment house in the real estate development built up on the grounds of the old race track, and we shared in the occupation of a roof garden from which we could see the snows of Popocatepetl and Iztaccihuatl. There was a young American couple occupying another apartment in the same building. The husband was also working at the Instituto, and we used to discuss together the book on prediction theory and control apparatus which I had promised Freymann.

I went to work very hard on this, but the first thing that puzzled me was what title to choose for the book and what name for the subject. I first looked for a Greek word signifying "messenger," but the only one I knew was *angelos*. This has in English the specific meaning "angel," a messenger of God. The word was thus pre-empted and would not give me the right context. Then I looked for an appropriate word from the field of control. The only word I could think of was the Greek word for steersman, *kubernētēs*. I decided that, as the word I was looking for was to be used in English, I ought to take advantage of the English pronunciation of the Greek, and hit on the name *cybernetics*. Later on, I found that a corresponding word had been used since the early nineteenth century in France by the physicist Ampère, in a sociological sense, but at that time I did not know it.

What recommended the term cybernetics to me was that it was the best word I could find to express the art and science of control over the whole range of fields in which this notion is applicable. Many years before, Vannevar Bush had suggested to me that new scientific tools should be found to deal with the new theories covering control and organization. Ultimately I began to look for these tools in the field of communication. My early work on probability theory, as exemplified in my studies of the Brownian motion, had convinced me that a significant idea of organization cannot be obtained in a world in which everything is necessary and nothing is contingent. Such a rigid world is organized only in the sense in which a rigidly welded bridge is organized. Everything depends on everything else and nothing on one part of the bridge structure rather than on another. The result is that in such a bridge there is no way to localize the strains, and unless a welded bridge is made of materials which can give and readjust their internal strains, the strains are almost certain to be so concentrated that at some place or other the bridge will break or tear and will collapse.

Thus, a bridge or a building is able to endure only because

it is not completely rigid. Similarly, an organization can exist only if the parts of it can give to a greater or less degree in response to the systems of internal stresses. Organization we must consider as something in which there is an interdependence between the several organized parts but in which this interdependence has degrees. Certain internal interdependencies must be more important than others, which is the same thing as saying that the internal interdependence is not complete, and that the determination of certain quantities of the system leaves others with the chance to vary. This variation from case to case is a statistical one, and nothing less than a statistical theory has enough freedom in it for the notion of organization to be significant.

I was driven back on the work of Willard Gibbs, and on the conception of the world not as an isolated phenomenon but as one of many possible phenomena with an allover probability distribution. I was forced to consider causality as something of which there can be either more or less rather than as something which is either there or absent.

The whole background of my ideas on cybernetics lies in the record of my earlier work. Because I was interested in the theory of communication, I was forced to consider the theory of information and, above all, that partial information which our knowledge of one part of a system gives us of the rest of it. Because I had studied harmonic analysis and had been aware that the problem of continuous spectra drives us back on the consideration of functions and curves too irregular to belong to the classical repertory of analysis, I formed a new respect for the irregular and a new concept of the essential irregularity of the universe. Because I had worked in the closest possible way with physicists and engineers, I knew that our data can never be precise. Because I had some contact with the complicated mechanism of the nervous system, I knew that the world about us is accessible only through a nervous system, and that our information concerning it is confined to what limited information the nervous system can transmit.

It is no coincidence that my first childish essay into philosophy, written when I was in high school and not yet eleven years old, was called "The Theory of Ignorance." Even at that time I was struck with the impossibility of originating a perfectly tight theory with the aid of so loose a mechanism as the human mind. And when I studied with Bertrand Russell, I could not bring myself to believe in the existence of a closed set of postulates for all logic, leaving no room for any arbitrariness in the system defined by them. Here, without the justification of their superb technique, I foresaw something of the critique of Russell which was later to be carried out by Gödel and his followers, who have given real grounds for the denial of the existence of any single closed logic following in a closed and rigid way from a body of stated rules.

To me, logic and learning and all mental activity have always been incomprehensible as a complete and closed picture and have been understandable only as a process by which man puts himself *en rapport* with his environment. It is the battle for learning which is significant, and not the victory. Every victory that is absolute is followed at once by the Twilight of the gods, in which the very concept of victory is dissolved in the moment of its attainment.

We are swimming upstream against a great torrent of disorganization, which tends to reduce everything to the heat-death of equilibrium and sameness described in the second law of thermodynamics. What Maxwell, Boltzmann, and Gibbs meant by this heat death in physics has a counterpart in the ethics of Kierkegaard, who pointed out that we live in a chaotic moral universe. In this, our main obligation is to establish arbitrary enclaves of order and system. These enclaves will not remain there indefinitely by any momentum of their own after we have once established them. Like the Red Queen, we cannot stay where we are without running as fast as we can.

We are not fighting for a definitive victory in the indefinite future. It is the greatest possible victory to be, to continue to be, and to have been. No defeat can deprive us of the success of having existed for some moment of time in a universe that seems indifferent to us.

This is no defeatism, it is rather a sense of tragedy in a world in which necessity is represented by an inevitable disappearance of differentiation. The declaration of our own nature and the attempt to build up an enclave of organization in the face of nature's overwhelming tendency to disorder is an insolence against the gods and the iron necessity that they impose. Here lies tragedy, but here lies glory too.

These were the ideas I wished to synthesize in my book on cybernetics. My first goals were rather concrete and limited. I wanted to give an account of the new information theory which was being developed by Shannon and myself, and of the new prediction theory which had its roots in the prewar work of Kolmogoroff and in my researches concerning anti-aircraft predictors. I wished to bring to the attention of a larger public than had been able to read my "yellow peril" the relations between these ideas and show it a new approach to communication engineering which would be primarily statistical. I also wished to alert this larger public to the long series of analogies between the human nervous system and the computation and control machine which had inspired the joint work of Rosenblueth and me. However, I could not undertake this multiform task without an intellectual inventory of my resources. It became clear to me almost at the very beginning that these new concepts of communication and control involved a new interpretation of man, of man's knowledge of the universe, and of society.

Communication is by no means confined to mankind, for it

is found in different degrees in the mammals, the birds, the ants, and the bees, to say the least; but, notwithstanding all the communication involved in the cries and nuptial dances of the birds, the dumb play by which a bee indicates to its hive mates the direction and distance of sources of honey, and all the rest of these modes of communication which we are just beginning to understand, man's language is more developed and more flexible than that of the animals and presents problems of a quite different sort.

Beside the obvious multiplicity of tongues and the wide scope of any individual language as a mode of expression, the extensive areas of the brain which seem to be devoted to the different aspects of speech and hearing, and reading and writing, testify to the overwhelming human importance of highly developed methods of communication.

To communicate with the outer world means to receive messages from it and to send messages to it. On the one hand, it means to observe, to experiment, and to learn; and on the other to exert our influence on the outer world so that our actions become purposeful and effective. Experimentation in fact is one form of a two-way conversation with the outer world, in which we use outgoing commands to determine the conditions of incoming observations, and in which, at the same time, we use our incoming observations to increase the effectiveness of our outgoing commands.

Communication is the cement of society. Society does not consist merely in a multiplicity of individuals, meeting only in personal strife and for the sake of procreation, but in an intimate interplay of these individuals in a larger organism. Society has a memory of its own, far more durable and far more varied than the memory of any individual belonging to it. In those societies which are fortunate enough to possess a good script, a large part of this communal tradition is in writing, but there are societies which, without writing, have preserved a whole tradition in the form of a technique of ritual memorization of tribal chants and histories.

Sociology and anthropology are primarily sciences of communication, and therefore fall under the general head of cybernetics. That particular branch of sociology which is known as economics and is distinguished by possessing rather better numerical measures of its values than the rest of sociology is a branch of cybernetics, by virtue of the cybernetic character of sociology itself. All these fields share in the general ideology of cybernetics, even if many of them are as yet insufficiently precise in their numerical techniques to make it worth-while to take advantage of the full mathematical apparatus of the larger subject.

Besides its function in these already existing sciences, cybernetics is bound to affect the philosophy of science itself, particularly in the fields of scientific method and epistomology, or the theory of knowledge. In the first place, the statistical point of view so manifest in cybernetics and in my earlier researches forces us to a new attitude toward order or regularity. Perfect information has nothing in it that is measurable, and measurable information cannot by that token be perfect. If we can measure degrees of causality (and much of my work on information theory has indicated that that is a perfectly possible goal), then that can only be because the universe is not a perfectly tight structure but one in which small variations are possible in different regions. We can then observe how much a change in one aspect of the universe will bring out changes in others.

Thus, from the point of view of cybernetics, the world is an organism, neither so tightly jointed that it cannot be changed in some aspects without losing all of its identity in all aspects nor so loosely jointed that any one thing can happen as readily as any other thing. It is a world which lacks both the rigidity of the Newtonian model of physics and the detail-less flexibility of a state of maximum entropy or heat death, in which nothing really new can ever happen. It is a world of Process, not one of a final dead equilibrium to which Process leads nor one determined in advance of all happenings, by a pre-established harmony such as that of Leibniz.

In such a world, knowledge is in its essence the process of knowing. There is no use in seeking for a final knowledge in an asymptotic state of the universe at the end of time, for this asymptotic state (if it exists) is in all likelihood timeless, knowledgeless, and meaningless. Knowledge is an aspect of life which must be interpreted while we are living, if it is to be interpreted at all. Life is the continual interplay between the individual and his environment rather than a way of existing under the form of eternity.

All this represents the manner in which I believe I have been able to add something positive to the pessimism of Kierkegaard and of those writers who have taken Kierkegaard as their inspiration. Among these, the most important are the existentialists. I have not replaced the gloom of existence by a philosophy which is optimistic in any Polyanna sense, but I have at least convinced myself of the compatibility of my premises, which are not far from those of existentialism, with a positive attitude toward the universe and toward our life in it.

These are the main ideas which I mulled over in my mind as I was writing my book on cybernetics. I discussed them with Arturo and with the American physiologist who was my neighbor in the same apartment house. We all felt hopeful that these ideas might amount to something, although none of us, not even I, had any concept of the excitement which they were actually to cause when they appeared in print.

I had some qualms about taking so much time away from the work Arturo and I had embarked on. This had languished slightly for a cause for which neither Arturo nor I was completely responsible. Arturo is an afternoon and evening worker who does not get steamed up until 3 or 4 P.M., and he can keep going until well after midnight. I am a morning worker, at my best on waking up, and I begin to lag by two in the afternoon, being totally unable to do creative work after dark. The result is that there were many gaps in our collaboration, which I could only fill by undertakings initiated by myself and which I did fill by the writing of *Cybernetics*.

In my work on this book, I was spurred on by certain fortuitous circumstances which were very threatening at the time and forced me to devote my energy to this new undertaking, which has become the basis of all my later career. Expenses were piling up on me from various sources, and I had no accumulated wealth from which to pay them off. I decided to do what many another writer has done; if it was possible, I would write myself out of this financial hole. To run ahead a little, this is exactly what I succeeded in doing; and though my writing has never tended to make me a rich man, *Cybernetics* represented the beginning of whatever security I enjoy at the present.

By now the second term of the M.I.T. academic year was approaching, and I was getting ready to return to the States. Very shortly before my return, I completed the book and sent it to Freymann, in Paris. There was a burden off my soul, and I spent some of the few days which remained to me in Mexico in visiting Taxco and in having a good time with my Mexican friends.

For some years I had been developing cataracts, and by this time they had advanced in both eyes to the point where they seriously interfered with my reading. There was nothing to do but have my lenses out. Naturally, an eye operation is an emotionally alarming thing. However, I was lucky enough to have an oculist in whom I had great confidence and who knew how to handle me emotionally as well. The result was that I found the operation less of an ordeal than I had expected and

I was quite ready for the operation on the other eye when the time came and for the series of minor operations on both eyes which were necessary to give me the largest measure of eyesight possible.

To a considerable extent, my nearsightedness and my cataract operations canceled each other. The final result is that I now have better distance sight without glasses than I have ever had and that my reading vision is of a tolerable quality. Nevertheless, the operations have left me with eyes which are rather sensitive both to excessive light and to long use. They have forced me to change my habits of work and in some respects to change them in a way very advantageous to myself.

I now do most of my mathematical work on a blackboard rather than on paper, and this relieves me from the unpleasant necessity of going from near vision to far vision and back again, which I can do only by the use of bifocals or trifocals. Even more, it has made me supplant my habits of writing either in longhand or on a typewriter by the effective use of very efficient secretaries.

The mere act of writing had always been an unwelcome task for a man of my physical clumsiness, and my antipathy toward it had tightened my style and introduced a general crabbedness into my literary work. Now I was free of this, and, since my operations, I have become far more of a literary man than I ever thought possible before.

I have always been conscious of literature as something at least as much for the ear as for the eye. One thing that contributed to this idea was the period of six months when, at the age of eight, I had been forbidden to use my eyes and all my instruction was by ear. Dictating forces on the author a consciousness of sound, and I like it very much. I have a memory considerably better than average, and the absence of notes is no handicap to me. When an idea occurs to me which must be pieced into a larger manuscript, I dictate it to my secretary, and we work together on smoothing the joints.

I do all of my dictating directly to a secretary, and I am repelled by the impersonality of the dictating machine. No secretary who is not an educated person and a person of taste could do the work which I require of her, and such a person can and should always give a continual criticism of what I am dictating by reaction and expression, if in no other manner. There is thus established what I should call, in my cybernetic vocabulary, a feedback process, which I use to full advantage.

Moreover, in my periods of dictating, there are long moments in which I am thinking of the next thing to say, and during which I can scarcely remember to stop the accursed cylinder of a dictating machine and then to start it up again.

I showed the manuscript of my cybernetics book to the M.I.T. authorities, in particular to the officials of the Technology Press. They were much interested. Indeed, they hoped that they might find a way to publish the book in America.

From one point of view this was not difficult, as the book was written in English, even though it was to appear in a French series. From another point of view, since I had already granted Freymann the rights in the book (for it was accepted as soon as it was received) there were both legal and moral considerations to be settled before we could use his sheets for the offset printing of an American edition.

All this was settled at last, and the Technology Press and John Wiley and Sons took the book over. By the way, the same collaboration of publishers was reprinting my "yellow peril" at the same time.

Freymann had not rated the commercial prospects of *Cybernetics* very highly—nor, as a matter of fact, had anybody on either side of the ocean. When it became a scientific bestseller we were all astonished, not least myself.

This book at once transformed me from a working scientist with a good but limited reputation in my field into something of a public figure. This has been gratifying, but it has had its disadvantages, for I have since been committed to work with the most varied scientific groups and I have had to form part of a movement which has rapidly grown beyond the possibilities of my personal control.

Cybernetics was a new exposition of matters about which I had never written authoritatively before and, at the same time, a miscellany of my ideas. It came out in a rather unsatisfactory form, as the proofreading was done at a period at which I could not use my eyes and the young assistants who were to have helped did not take their responsibility seriously.

After the publication of the book, which got good reviews and, as I have said, a most unexpected sale, I came to be in much demand for the writing of papers of a more or less popular nature and for public lectures. For a time I followed the course of acceding to the blandishments of invitations to write and to speak, which gave me a new and perhaps spurious sense of importance.

Since then I have learned that if I want to contribute anything more to science, and if I wish to keep in reasonably good health, I must conserve my energies. In general, invitations for lecturing do not return to me in rewards or in prestige what they take out of me in fatigue. I have also found out by bitter experience how much a lecturer must protect himself against exploitation.

For very similar reasons, I do not touch requests for engineering consultation. In a field like my own, the consultant is usually far more interested in getting my name than in getting any ideas which I may have. Moreover, the ordeal of being quizzed by relays of company engineers, of contacting and meeting a number of strangers who are primarily interested in draining me dry, and of remaining affable under all these exactions, is something which the torturors of the Inquisition omitted to add to their repertory.

During all this time my daughters were in their late teens and early twenties, and going through college. Barbara remained undecided for some years between a scientific career and journalism. She had first spent a year at Radcliffe, and then studied at M.I.T. for a time. She did her journalistic work at Boston University, but it was not until after her marriage to Gordon Raisbeck that she completed these studies at Drew University, near her home in Morristown, New Jersey. In the time between, she had done a considerable amount of work in scientific journalism for Science Service in Washington.

Peggy matriculated at Tufts College, which was the school where I had done my undergraduate work. She specialized in biochemistry, and after her graduation and a certain amount of graduate study at M.I.T., in London, and at Boston University. She worked for a while at the Worcester Foundation for Experimental Biology. Peggy married shortly thereafter and is now active with a pharmaceutical firm in northern New Jersey.

Both my sons-in-law are engineers in the Bell Telephone Laboratories, doing work of a definitely mathematical nature as well as other more applied work. Thus, I have in my own family exemplified that peculiar genetics of mathematics of which I have already spoken, that mathematical ability goes down from father-in-law to son-in-law.

As 1950 came on, I received an invitation to go to France as a Fulbright fellow and to lecture at the Collège de France. On the French side, this offer emanated from Mandelbrojt. I finally decided that I could not spend a whole year on this undertaking, and I did not sail for France until December.

My French friends found me a hotel in Savoy where I could rest up before my rather arduous duties began. These had become arduous because I was also to participate in a congress on high-speed computing machines and automatization which was to take place in Paris early in January, 1951.

After this congress was over, I went to England for a few weeks and stayed with the Haldanes. Soon Margaret came over together with Peggy.

Margaret and I left immediately for Paris. We were put up for a few weeks in a building belonging to the Paris Observatory and at once found ourselves taken up into the intellectual and social life of the community.

I enjoyed my teaching at the Collège de France and was treated exactly as one of their own men. Every day of my lectures, of which there were twenty, I would go to the little office, reflect over my talk for a few minutes, sign the daybook, and be conducted into the lecture room by the wooden-legged *appariteur,* or university servant. I lectured in French, calling on my audience for help when my vocabulary ran out.

The very first day, I found an old friend at my lectures. It was a French doctor who had been working at the Instituto Nacional de Cardiologia in Mexico, and who had taken care of me on one occasion when I had overexerted myself. He saw to my health through our whole stay in France and invited my wife and me to many very delightful occasions at his apartment. He has visited America several times since, and we have had the chance to return his hospitality on our own soil.

The mathematicians took us completely into their family life. I saw a lot of dear old Hadamard and his wife, both of whom seemed to us almost ageless, although they were in their eighties. Fréchet and Bouligand were also our hosts.

I gave several outside lectures, some of them on engineering topics before a group from the Ecole Supérieure de Télécommunications. I also gave a philosophical lecture in a hall almost under the shadow of the home of Existentialism; the apartment of Sartre. We went to a salon at the apartment of one of the philosophy professors, where I was lionized in the thorough French manner.

I spent much time gossiping with Freymann in his back office, playing chess in the Bar Select on Boulevard Montparnasse, and in the other minor amusements of the city. We were great frequenters of the movies, and we learned to know a little bit about the good restaurants and cafés of Paris.

I had written another book for general consumption the previous year. It was a more popular account of cybernetics, with the social elements underlined. This was *The Human Use of Human Beings,* which had been published by Houghton Mifflin and has appeared as a paperback among the Anchor Books of Doubleday. I now tried to vend it to a Paris publisher for translation into French. I finally found such a publisher in M. Dufèze, of the Editions des Deux Rives.

Peggy visited us at Easter, and the family went on a trip to Nancy, where I gave a talk. Laurent Schwartz and his friends were as cordial as they had been on the previous occasion. My wife's French is better than mine, and Peggy found herself quite able to take part in our conversation and in our social life.

Later on in the spring, when my lectures in Paris had come to an end, Margaret and I went to Madrid. The summer before this last French trip, there had been a meeting of the world Mathematical Congress at Cambridge, Massachusetts, in which I had participated. At that time, the Spanish were very cordial to me, and I had received an invitation to lecture in Madrid. I protested that they might not like my views when they found out what they were, but they pooh-poohed the matter.

I accepted their invitation. My host had meanwhile read some of my writings and had decided that my views were dangerously liberal for me to present in a totalitarian country. Although I speak Spanish as well as I do French, he requested me to give my lectures in French; as I am now convinced, so that they should be less generally understood. He enjoined me to speak of engineering and mathematics only, and not to speak of anything political, philosophical, or biological.

We were put up at an excellent hotel and received the most splendid hospitality, but throughout the whole period we felt that we were being isolated from the country and kept in ignorance of what was going on. In view of our knowledge of Spanish and our experience as travelers, this ignorance could not be maintained, for I used to go out for walks in the nearby park and talk with the people there, and Margaret and I made a train trip to the Escorial. When our host found out that we had thus eluded his surveillance he was very angry, and he was even angrier when another Spanish friend offered to take us on a motor trip to Seville.

Thus we were glad to get out of the country and return to the free life of France. We spent part of our vacation in the delightful Basque town of St.-Juan-de-Luz. There I turned myself in all earnest to a task which I had begun in America and continued in Paris and Madrid—the writing of the earlier volume of this autobiography, which appeared under the title of *Ex-Prodigy*. It was a tremendous emotional strain to relive the severe experiences of my childhood as an infant prodigy, but the writing of them was also the best sort of psychiatric therapy.

We found some difficulties in securing transportation home, and we returned to Paris to clear up this matter, together with other questions of importance. We settled in a delightful left-bank hotel near the church of St. Germain. Then we went back to Savoy to rest until we were ready to return to the States.

One of the doctors of the Savoyard town where we stayed, was the father of our Paris medical friend. Toward the end of our stay in Savoy, my overexertion in lecturing and in writing sent me to bed with a racking headache, and for a while I had to go to the Geneva Canton Hospital. Meanwhile, my Paris friend had written to his father suggesting a suitable treatment, and with the aid of this I was soon much relieved. Even so, our trip to the boat at Genoa was a great deal of torture, and I was handed over to the care of the ship's surgeon. He continued the treatment, and by the time I reached home I was a reasonably well man but dead tired.

My wife and I went almost immediately to Mexico, where the university was celebrating its four hundredth anniversary. It was a time when honorary degrees were given out, and I received one of them. Mexican festivities are lavish in the extreme, and the two weeks of ceremonies, while delightful, exhausted me. I continued my work with Arturo until we returned to the States, in January 1952.

Even before this, Indian mathematicians had begun to negotiate with me to have me visit India and lecture there. It was not until the Christmas holidays of 1953 that I felt that I had any right to go.

16

India.

1953

In December 1953, I started on a seven weeks' lecture trip in India as a guest of the Indian government and a group of government-supported institutions. The occasion for the trip was the All-India Science Congress at Hyderabad. A trip of this sort had long been under discussion. It had hung fire for some years because it was difficult to make the several demands of the Indian Government and my own meet one another precisely, not only in time but also in the nature of the trip itself. I was disinclined to take the trip alone without the protection and support of my dear wife; but this was manifestly impossible to ask for if my stay were to be brief, nor was it really practicable at the time for me to suggest a longer stay.

However, it became clear fairly early in 1953 that matters were going to break just right for something like a six or seven weeks' trip to India. My health, which had been rather strained by the severe ordeal of my recent successes, had recovered sufficiently to make it clear that with a bit of judicious care I should be equal to the trip. Moreover, my daughter Peggy was obviously on the point of becoming engaged to a young engineer and college friend, John Blake. In fact, the engagement was announced early in the fall term of 1953. On the one hand, this made it essential that my wife remain in the United States to take care of the manifold preparations for a wedding. On the other, it caused a lively period of preparations, during which I should be for a time more or less of a superfluity in my own household. I closed the arrangements and left the United States on the nineteenth of December, to return a week before my daughter's wedding, set for the twentieth of February.

As in the case of my earlier experiences in China and Mexico, so in my Indian trip my motive was more than restlessness or idle curiosity. More and more Indian authors are publishing in our scientific journals, and we need the Orient more and more to supplement a West which is showing the intellectual and moral enfeeblement following two World Wars. I was glad to have the opportunity to see something of yet another new recruit to international scientific life and to get the feel of its atmosphere.

I had, of course, been prepared somewhat for my Indian voyage by my earlier trips abroad. My visits to China and Japan had given me a degree of insight into the Orient and into the special problems of countries combining great intellectual ability with great poverty and now just beginning to enter upon the stage of a truly international scientific life. On the other hand, Mexico, with which I have become thoroughly familiar in the course of a decade, combines some of these problems with those belonging to a tropical climate. I had known many Indian students and colleagues both in the United States and in England. These people, many of whom I was to see again in India, had given me a certain picture of specifically Indian matters, and in particular of the strong religious attitudes which underlie Indian life.

I left Boston for Paris by air on the afternoon of December nineteenth. I am not an enthusiast for air travel. The trip is too short to encourage new contacts or to permit one to undergo the spiritual preparation for new and intense experiences.

There was a strike on at the Paris airports, and we were not sure of our destination. In fact, it was not until we had made an unscheduled landing at Shannon that we learned that we were to leave the plane at Brussels. I finally learned that we would be forwarded to Paris by a chartered bus, and not by plane. We drove hour by hour through Belgium and crossed the frontier well along in the day. Arriving after dark at the Gare des Invalides, in Paris, I found that my further trip by Air India was still unsettled because of the strike.

The three days which I spent in Paris as a result were a delightful combination of visits to friends, impromptu lectures, and contacts with my publishers and my colleagues. I had the good luck to find that there was a small sum waiting for me at Hermann et Cie. in the way of royalties I did not even know existed. But poor Freymann, who broke this good news to me, died of a stroke soon after my visit.

I found that I would have to leave Paris on the night of the twenty-third for Geneva, from which my plane was to leave for India. I spent Christmas Eve in Geneva with a neurologist whom I had seen many times before and whom the family knew very well.

To step upon a plane of Air India was to enter India in Switzerland. The pilots were Indians—largely Parsis—as were the two hostesses. The crew contained the extra contingent of servants which one must always expect in the Orient, and the meals were specially prepared to suit the religious and dietary demands of the passengers. Since I had announced myself a vegetarian, I found that this was an ideal arrangement for me.

We landed in Bombay, where I was rushed through customs and immigration inspection with the help of the secretary of the Indian Atomic Energy Commission, who had been deputed to take care of me and the other foreign scientific visitors in Bombay who were arriving for the All-India Science Congress. We hit it off from the beginning, and he took me out to the beach at Juhu to have tea with him and his wife under the palm trees. The cordial reception which I found upon first putting foot on Indian soil continued throughout my whole stay in the country.

After settling down at the Taj Mahal Hotel, which is a fascinating combination of the East and the West, I went the next day to the cornerstone laying of the new Atomic Energy Institute on military territory near the harbor. There was an interesting group of notables present, including Nehru himself, who gave a short and excellent speech. Among the visitors present was the Cardinal of India, a tall Goanese gentleman representing the old Portuguese religious tradition in India. This Christian tradition, like the old Syrian Christian tradition of the South, is often ignored by foreigners or, at any rate, not sufficiently emphasized. Actually, the Portuguese are older in India than the Mogul emperors. Notwithstanding the fact that Goa is not at present in India itself (as of the date of my writing, at least), Goanese are found all over India, particularly on the Bombay side. They consider themselves a thoroughly Indian element in the population. I had the pleasure of meeting Goanese officers both in the Army and in the Navy, and it was manifestly clear that they considered themselves, and were regarded by the others, as true Indians.

Another thing that struck me was that the cornerstone ceremonies were in English. English remains one of the chief languages of India, even though there is a deep-seated movement of the politicians to replace it by either Hindi or the local languages, and to confine a thoroughly English education in India to the people who, in part at least, are of English origin.

These people, the Anglo-Indians, are an older and more important element in the population than many might be inclined to expect, and they are having rather a thin time under the pressure to accept themselves as Indians and to be accepted by other Indians. In fact, there is no doubt that this is the only real future for them.

Anglo-Indian ladies, together with Parsis and southern Christians, serve as air-line hostesses for the very peculiar reason that this job involves the wearing of European dress, and European dress for women is looked upon with very little favor by both Hindus and Moslems.

For all officer ranks in the Army and Air Force, and to a certain extent for all ranks in the Navy, English is the only acceptable language. This results largely from the technical nature of modern warfare and from the lack of adequate technical vocabulary and phraseology in any of the native languages, which indeed borrow heavily from English in these matters. English remains the language of Indian science in all its branches, and although there is an effort to switch over to Hindi for the future, it is still too early to say how successfully this move will be.

The English rule in India has lasted about as long as the Mogul rule before it, and its effects were not transitory. It is of course fashionable for patriotic Indians to play up their own history rather than that of the English, and even to write of the Indian Mutiny as a war for independence. Nevertheless, the deep hatred for England which characterized much of the first half of this century has largely subsided, and there is probably no foreign country that is so well thought of by India as England, now that respect for England has ceased to be a sign of subservience to an alien rule.

The Englishmen who have helped India through its transition to a full independence, and who remain on in India in positions for which no adequate Indian replacement has yet been found, are regarded not only with respect but also with affection. I am referring to such people as Lord and Lady Mountbatten, to the high army officers who have stayed on, on loan in technical jobs, and to a certain small group of heads of scientific institutions.

What I am saying here goes for all of India, but I was particularly struck by the role of the English language in the South. I have been told by Indian friends who are not professional Anglophiles that in the city of Madras, ninety-five per cent of all classes of the population understand English and speak it tolerably. This is probably due to the fact that Tamil is a very difficult language, and that it does not pass current outside its own region in India, and that Hindi is quite as foreign and quite as difficult for a Tamil as English can be.

One finds Tamil-speaking men in all parts of India who, because of their native ability and intellectual discipline, are more than likely to use English as the habitual vehicle of their communication with their north Indian friends.

After the laying of the cornerstone, I went to a party at which Nehru was the chief guest and had a chance to meet him and to see the enormous respect and even reverence with which he is regarded. He seemed rather tired and frail, but I hear that he has great resources of strength. These India and he will need; for in the Congress Party, of which he is the leader, there appears to be no man of even approximately his stature as a second-in-command.

The day after the cornerstone laying I went by plane to Ahmadabad. I was sent there to participate in the meeting of the Indian Academy of Sciences. Here the chief figure was Sir C. V. Raman, the physicist and Nobel laureate.

I was a house guest of Professor Vikram Sarabhai, the physicist, who turned out to be a close friend of my friends the Vallartas. In fact, the Vallartas had been house guests there shortly before me. Mme. Sarabhai is a well-known dancer in the classical Indian style, and she maintains her troupe of dancers and musicians in the house. I had seen her perform with them in Mexico at the occasion of the four hundredth anniversary of the university.

In view of our having friends in common, the Sarabhais took me literally into their family. In fact, I was invited to participate in the weekly dinner of the larger family, which took place at the house of Vikram Sarabhai's father, a leader in the textile industry in Ahmadabad.

At the family feast—I should say clan feast, because the Indian family is a more comprehensive group than the Western family—I was the only non-Hindu present and, as an indulgence to my foreignness, I was granted the use of a table at which to eat, while the rest ate off little platforms on the floor.

Among the members of the Sarabhai family present, perhaps the most interesting was Vikram's maternal grandfather, an old gentleman who had served many native states as a dewan, or prime minister. Slender, ascetic, and unbelievably aristocratic-looking, such old men as Vikram's grandfather, fill an important role in the development of the new India. The Vedic injunctions lay down very precisely the proper course of a man's life. According to my understanding of them, which I am not sure is literally authentic, a man should spend twenty years as a youth, twenty years as a soldier (or, perhaps more precisely, as an active participant in the struggles and controversies of the world), and twenty years as a head of a family— that is, of the larger family in the Indian sense. After this, it was recommended that he become a *sunnyasi*, or religious recluse, and that he devote the few years left to him to the contemplation of the divine and to the attainment of that very Indian sort of salvation which is known as nirvana. In this way, he may interrupt the ordained sequence of reincarnation into new animal and human bodies.

The classical type of *sunnyasi* still exists in India, and the story of one of these is given in Kipling's tale "The Miracle of Purun Bhagat" in the *Second Jungle Book*.

Nevertheless, the purely contemplative life is not enough for an India which is mistress of its own fate, and which needs to have so much done to interrupt the continued succession of poverty, ignorance, and misery that it cannot spare its able and experienced men the luxury of securing their own salvation. Thus, the deep and religious impulse of these fine old men to the otherworldly has been sublimated into a selfless service of the needs of the community, in those matters in which experience and integrity are indispensable, and in which there is no personal motive of advantage.

The Indians are not slow to find in their own scriptures a justification for this more worldly and practical but equally religious equivalent for the life of the *sunnyasi*. In doing this they are on a perfectly sound basis. No country can make adequate use of motives and modes of action which are merely passed on to it from the outside, but must find somewhere in its own body of tradition and in its own soul the moral sanction for the developments which are necessary to meet new problems.

Like other modern people, some of these old gentlemen wear Western costume, but there are many who cling to the costume and ways of their own country. When they do so, they exhibit an astonishing grace and beauty. It is remarkable how aristocratic a simple wool shawl can look when it falls over the shoulders of a beautiful and gentle old sage, in the lines of the garments of a Buddha in one of the temples of Ajanta or Ellora.

At the Sarabhai family dinner, Sir C. V. Raman and Lady Raman were present. Raman is both an important individual and a type significant in the new development of India. He is a southern Brahmin with a subtlety and depth of intellectual power which belong to many southern Brahmins as their birthright. However, there is in his character an additional positiveness and definiteness which show a familiarity with authority and a readiness to take an active part in life. Raman is an applied physicist and a good experimenter rather than the sort of theoretical physicist one might expect in India.

I gave a couple of lectures at Ahmadabad, one on my attempts to do something in quantum theory and one on the theory of prediction. I then went on to Poona, where I visited the university, which is headed by one of these fine old Indian gentlemen of whom I have spoken. I also talked at the National Chemical Laboratory. I met students there who were working on diverse and important problems in physical chemistry. The head of this laboratory, G. I. Finch, is the great climber of Mt. Everest, who is extremely popular among his subordinates and among Indians generally.

India has made a right-about-face in its attitude to the English. It had been difficult to make friends across the barrier between the rulers and the ruled. The ruler may be a snob, and the subject may fear that he is playing the sycophant and identifying himself with authority for his own advantage. The separation of India from England has greatly facilitated true Anglo-Indian friendship.

The wise and moderate counsel of Gandhi and Nehru has contributed much to making England the most popular western country in India. Nor must we forget the intelligent policy followed by the Mountbattens, as the last viceroy and vice-reine, in relinquishing British authority in India and in making ready for the new order. I will not say that there does not remain much criticism of the British policy in the past, but this is combined with the healthy realization that such criticism is a very poor refuge from facing the living problems of an old-new country.

After my return to Bombay, a Russian scientific delegation came to India to participate in the Indian Science Congress that was to take place in Hyderabad. Obviously, one of their purposes was to perform indirect propaganda by showing the Indians the cream of Soviet science. On the whole, the members of this delegation were well-chosen, for while it contained a few of those who were principally political apologists, it also contained a number of sincere and unpolitical scientists of high rank and fine personality.

Throughout my trip I was to be lodged at the same hotels at which the Russians were placed. It was necessary for me to define my policy from the start. I thought, and I believe rightly so, that any display of personal hostility was unjustified and would not redound to the good name of America among the Indians. I went to several members of the delegation and said, "Look here, we're to be together for some weeks, and I don't want to embarrass you or myself. Let's be friendly and talk freely over such scientific matters as have no technical or political implications." I found a ready response to my attitude, and at no time in the trip did the real scientists of the delegation embarrass me by anything that could be propaganda, or any digging for information.

At the beginning, the Russian group consisted entirely of the people who had been picked as scientists. Of course, some of these scientists were in the fields of philosophy of economics, which naturally meant that they were expected to take an orthodox Marxian point of view. The remaining delegates seemed very much like scientists anywhere else and did not seem to be held to any sharp party line on their respective scientific problems.

At this early period, conversation with the Russian group

was relatively easy. Later on, they accumulated a body of translators and the like from the Bombay consulate and from their embassy at Delhi. This body was, in fact, of comparable size with the original delegation, and consisted of people who did not make on me and on the Indians nearly so favorable an impression as the scientific delegates themselves. They appeared rather to be policemen whose main purpose was to shepherd their charges away from possible western European and American influences and to prevent them from saying or hearing anything that might have been injurious to their government.

Scientists who had been conversing with me in very usable and effective English were forced by these channelers to speak almost nothing but Russian. Then the entire atmosphere of the meeting became, I shall not say hostile, but distinctly less open and friendly. Every time one of us fell into conversation with one of the Russian group, there was an interpreter sitting on a chair near by, either translating from the Russian, or at least taking in every word that was said in English. Occasionally they seemed to give the high sign to their charges that further conversation was out of bounds.

Thus, the Russians tended to withdraw themselves into a compact mass, dining together and having very little contact either with the foreign delegates or with those Indians who did not manifestly form an organized group of Soviet sympathizers. Here I relate what I saw, although I was told that towards the end of the meeting the Russians did in fact secure some more general and independent contacts.

Shortly after the Russians came, we moved on to Hyderabad, where all the foreign delegates were given quarters in the Hill Fort, a rather shabby and run-down palace which had formerly belonged to a son of the nizam. Here some of the vast rooms were turned into dormitory spaces, and I found myself sharing one of these with a couple of elderly and distinguished

English scientists. After a certain amount of feeling one another out, we got along very well together.

The isolation of the Hill Fort was most suggestive of the setting for an Agatha Christie detective story. Here we had together under one roof, and dining at the same table, a delegation from Russia and a mixed lot of other scientists like myself. Indian scientific notables and cabinet ministers came and went every few hours, but except for this we were quite alone, and none of us spoke a word of any Indian language. We gradually became a little fed up with the schism which divided us into two groups between whom there was a minimum of communication, and we westerners decided that the non-Russians should sit in alternate chairs at the table and force the Russians to take places in between us.

The day came when we non-Russians set out on a chartered plane for Aurangabad, from which we took a bus to the marvelous cave temples and sculptures of Ajanta and Ellora. Whatever the English have done in India, they have conserved its antiquities and historical monuments. The Indians are continuing in the good tradition which they found when they came to power by continuing to preserve these monuments and by encouraging their citizens to learn something about their own great past.

We spent a couple of nights in Aurangabad and left one morning for Hyderabad in a plane which we had expected to bring the Russians, who also were to visit the caves. There had been a series of mishaps in their setting out, and not only did they cancel their flight, but they remained plane-shy for the rest of their stay in India. Their decisions to discontinue the Ajanta trip were made rather thoughtlessly, for the Indians had put themselves out to find the Russians accommodations at Aurangabad.

We teased the Russians more than once, and I must admit they were good sports about it. When there was a meeting of

Friends of China at Hyderabad, they were actively prominent,
but I found that none of the Russians knew a word of Chinese.
I displayed my knowledge of Chinese a bit ostentatiously, and
I ribbed the Russians on their lack of knowledge of such an
important world language. Furthermore, I asked them why
they did not break ranks and make closer contact with the
people of the country.

I did this with the knowledge that a recommendation from
a westerner like myself would have a decided negative value
for them. In fact it was clear that their policy of sending delegates *en masse* was a weak one. Many of the westerners were
rich in Indian contacts and could really meet the Indians at
their homes.

At the end of our stay in Hyderabad, some of us were invited to a series of informal parties at which members of the
cabinet of Hyderabad were present, and to some at which they
were hosts. It was most heartening to see Moslems and Hindus
working together without discrimination of creed and to find
Hindu ministers at a breakfast table presided over by the wife
of a Moslem minister.

I took the plane from Hyderabad to Madras, where I was
met by my old friend, Vijayaraghavan, whom I had known
some eighteen years ago in England and later in America,
where he had visited my family. At the time he was a slender
young man concealing his Brahminical topknot in a spotless
white turban, which my two young daughters played with. He
left a deep impression on them, and I believe in fact that they
named a doll after him. He visited us again in 1952, and this
time he did not wear a turban, for there was no longer any topknot to conceal. My grown daughters regarded him with the
same affection which they had shown him as children.

In Madras he was my kind friend, host, and counselor; and
though I stayed at the hotel, not at his house, we spent the
entire waking time of my stay together and I dined several

times at his house. When one realizes that I am by Hindu standards a mle *chchha,* an outcast, and that a generation ago any Brahmin would have considered himself polluted by my very presence at meals, I felt that this was a very considerable favor and token of friendship. We used to go in the hours of dawn for a swim in the magnificent surf of the Indian Ocean, and he would bring along his daughter and his little grandson.

I gave a scientific lecture at his institute and met his charming group of friends. I received a splendid idea of the lively and cordial intellectual life of Madras. I lectured to a group of Vijayaraghavan's friends on the automatic factory and its possible effect on the future India. I think I was almost the only one there who wore European attire.

Finally, Vijayaraghavan, together with his mother and his daughter, went with me to a little cloth shop by a temple in the Mylapore suburb of Madras and helped me pick out for my daughter Peggy a splendid saffron silk sari with dark-red gold-woven trimmings, together with the material for a blouse to go with the sari.

On many occasions during my stay, we talked over many things both scientific and personal, and we speculated much on the lives that our grandchildren might live and whether they might not find a better world in which religious and racial prejudices should have abated and in which all the peoples could meet for all purposes in an atmosphere of universal humanity.

From Madras I made a brief excursion to the delightful city of Bangalore, where I saw more of Raman and I took an active part in lecturing and in the intellectual life. Thence I returned for a week as a guest of the Tata Institute at Bombay. Bombay was full of first-rate scientists, both Indian and foreign, and I found a rich opportunity both to teach and to learn and, in particular, to collaborate with and to criticize the work of several young mathematicians.

My special crony during this episode was Professor Kosambi, who had been a boy at the Cambridge High School during a period when his father, a refugee from British India, was working over the rich Sanskrit material of the Harvard Library. The son, perhaps owing to his American early training, has been a bit more of a fighter and a bit less of a serene Indian scholar than most of his fellow countrymen. However, I found that he was not the only Indian to counter my admiration for the serenity of the Indian soul with an equal admiration for the drive of the westerner.

Among the other Bombay scientists whom I met were Masani and Chandrasekharan. Masani is a Parsi, and I saw a good deal as well of his Parsi colleague Bhabha, of the Tata family. I found the Parsis an extremely interesting group, accepting the new India very thoroughly. Yet they were partly divided in soul between their patriotism and the quasi-European position, almost as westerners, which their small minority of a hundred thousand souls had accepted under the English.

My stay in Bombay was one of the most profitable parts of my Indian trip, and I shared my very newest work with my Indian colleagues. When I went to India I had already been at work on the problem of prediction of multiple time series such as, for example, the weather at two or more points. This led to a certain formal mathematical problem in factoring what are known as matrices. I thought that I had a complete solution of the problem already, but when I spoke to Masani, he showed me that the question should be conceived in a larger way than that in which I had conceived it and that there remained much to be done. While in Bombay, I turned my maximal effort toward the solution of the problem and was luckily able to close the books on it.

I think that the fact that I was engaged in active new creative work in India, rather than the mere presentation of work already done, brought me much closer to the Indian mathematicians that I otherwise could have hoped to come. At any rate, I tried to live up to my opinion that the best and indeed the only way of teaching advanced students in science is to participate with them in a common enterprise.

I gave a talk to the Bombay Rotary Club on national and racial relations, and I met an interesting assembly of former M.I.T. men, who seemed to be taking a very active share in the new national development of India. I also visited the College of St. Francis Xavier, where the Spanish Jesuit fathers seemed a much-beloved group, fraternizing freely with their Hindu, Parsi, and Moslem students. Indeed, I found the Church in India a much freer and more liberal institution than it had seemed to me in the Spain from which these fathers came.

From Bombay I went by airplane to Calcutta, where I had the privilege of working for a week at the Indian Institute of Statistics, headed by Professor Mahalanobis. The Mahalanobises received me as a house guest, and I was admitted to a charming degree of intimacy with the family. I was given a room which had been intended for the old age of their friend, Rabindranath Tagore.

There was a most interesting group of Indian and foreign scientists going and coming, and I profited by the wise and understanding criticism which Professor Bose, of the University of Calcutta, made of the new physical ideas of Armand Siegel and myself. I gave a number of lectures to the staff of the Institute, and I kept myself at their disposal for discussions of their own research.

I used to go to a nearby temple—the one that forms the scene of some episodes in the film *The River*—to think over my scientific work and I had the pleasure of being received cordially by one of the devotees there. He was a Post Office official who came every day to worship. He was bearded and dressed in a scrupulously neat Indian costume. He invited me to see the inner precincts of the temple, which, until recently, had been closed to non-Hindus.

The Mahalanobises and the charming company that gathered at their house talked over world science and world politics with me with a great freedom and directness. They sent me to see the sights of Calcutta, including the zoological gardens and the art museum.

I followed up my trip to Calcutta with a visit to Benares and later to Agra. Benares struck me as a sinister fairyland. Agra, on the other hand with its palaces and tombs, including the Taj Mahal, was a lesson on the possibility of combining kingly lavishness with discipline and proportion.

From Agra I went on to Delhi. There I saw the fine National Physical Institute run by Professor Krishnan, and I was much gratified to see the equal emphasis on scientific progress and on the setting-up of a corps of workmen to make this scientific progress possible. The work that was being done on the use of solar energy was already beginning to yield results, and promises even greater ones for the future. If it is merely confined to cooking by solar heat instead of over flames of burning cow dung, it will already have made an important contribution towards the improvement of the fertility of the Indian soil.

I lectured at Krishnan's institute as well as at the university. At the university, I took up the theme on which I had already spoken during my popular lecture at Mylapore, of the significance of the automatic factory for the future of India.

My stay in India led me to reflect on the future role of the country in an industrialized and scientific world. As I have said, Indian scientists are the intellectual equals of those in any country. On the other hand, the class of skilled technicians, the non-commissioned officers of science and technology, are much more difficult to recruit. In artistic matters Indian craftsmanship is excellent, but it tends to lack the precision and uniformity demanded by the workshops of the West. Much brave work is being done in recruiting a cadre of these non-commissioned officers, very largely within the military services themselves and in the great new national laboratories. The National Physical Institute depended for its supply of skilled workmen on the Sikhs, who show the same abilities in the workshop which have made them in the past one of the mainstays of the Indian army. However, the facilities of the country up to the present have rendered the new class of skilled workmen into which they are entering a rather limited one.

At the bottom of the population there is an unlimited supply of unskilled and not too efficient labor, which makes a country very susceptible to a devastating proletarianization of even worse character than that which took place in England under the early days of the industrial revolution.

In view of these circumstances, I doubt whether India should undertake its industrialization in accordance with the accepted western pattern of mass factory labor. This is one of the quickest roads towards an immediate industralization, and it gives India a chance to capitalize on its unquestioned asset of mass population. But I doubt whether this process is worth the price in human misery. Wretched and undernourished as a villager is, the industrial city promises to be even more wretched and to deprive the urbanized villager of whatever very small status he may have under Indian conditions of poverty. The unchecked growth of a nineteenth-century factory system is already making the outskirts of the great city into an unlovely hybrid of Indian famine and Manchester drabness.

I am not willing to ignore the possibility that the future industrialization of India may bypass much of the drabness and misery of Manchester or Chicago through the early introduction of the automatic factory. Misery is a result of unemployment, but it is even more a result of the sheer lack of goods. The automatic factory makes its demands on human efforts not at the bottom but at the very high level of the scientist-engineer and at the relatively high level of the small group of highly skilled trouble shooters and maintenance workers. It is quite in the cards that India can supply both of these within a matter of decades, while it can not supply a large group of fairly skilled factory workers able to earn enough to maintain them in a half-decent life for a large part of a century.

Of course, I may be wrong. The hothouse atmosphere of rapid industrial growth under the regime of the automatic factory may conceivably foster evils greater than any which it can alleviate. I do not know. What I do know is that the introduction of the new economics of the automatic factory may take place in India faster than most of us are willing to admit and that it might well be an easier avenue towards a prosperous and effective industrialized country than any of its alternatives.

In other words, this is a possibility which we indeed may have to discard, but which India cannot afford to discard without a thorough consideration of what it means. I am told that Nehru is interested in thinking out the possibilities of this alternative path to industrialization.

17

Epilogue

I am writing the last pages of this book at the age of sixty, an advanced period in the life of a creative mathematician. However, I am still at work, and I would not like to think that my efforts are now over. Many of my ideas still contribute to the growth of engineering and physics, so that a book of this sort can be nothing more than an interim report.

Many scholars find it interesting to speculate whether their motives for entering scholarship and their later success are due more to heredity or to environment. In my own case, it is particularly difficult to separate these two factors, because, to a large extent, my heredity was my environment.

I owe to my father not only his share of those genes which I carry but the sort of training which he conceived to be proper for a boy with exactly those traits of character which I derived

357

from him. Without my share in Father's nature, I would have been an unfit subject for his training; and without his training, the potentialities which I derived from him might well have gone undisciplined and unrealized.

One part of Father's own outlook and one part of the training which he gave me was a very thorough synthesis of the theoretical and the practical. Father was a philologist who regarded the history of languages not as the quasi-biologic growth of almost isolated organisms but rather as an interplay of historic forces. For him, philology was a tool of the cultural historian, exactly as the spade is of the archeologist. It is not surprising that the son of a father who could not be contented with the formal and the abstract in the study of languages should himself fail to be contented with that thin view of mathematics which characterizes those mathematicians who have not made a real contact with physics.

My father was one of the most independent of men, and I could never have been a loyal son of his without declaring my own independence, even of himself. Fundamentally, he did the research he liked to do, and that was the kind of work which I had seen about me from my earliest childhood. His work was disciplined not in accordance with the injunction of others but in accordance with the inner demands of a stern and self-critical nature. As my father's son, I could do nothing else but proceed in this pattern.

I have done those things I have done not in response to orders from outside but because my wishes followed a pattern which appealed to me and because the individual pieces of work which I have done have seemed to build up in a definite organized direction. My discipline has been a self-discipline, in the image of the discipline imposed upon me as a child by my father.

The discipline of the scholar is a consecration to the pursuit of the truth. It involves a willingness to undergo such real sacrifices as are demanded by this consecration, whether they are sacrifices of money or sacrifices of prestige, or even in the extreme (but not unprecedented) case, of personal safety. However, the main part of this discipline is intrinsic and belongs to one's relation to science itself rather than to one's reaction to the external environment within which science is carried on.

In the first place, discipline does not preclude the making of errors. What it does preclude is the retention of an error which has clearly and distinctly betrayed its wrongness. If a theorem is inconsistent, or if a proof refuses to become complete under the greatest pressure which you can exert, cast it from you.

This is the negative side of intellectual discipline. There is, however, a positive correlate to this. If a theorem merely looks grotesque or unusual and if your maximum effort cannot discover any contradiction, do not cast it aside. If the only thing that seems to be wrong about a proof is its unconventionality, then dare to accept it, unconventionality and all. Have the courage of your beliefs—because if you don't you will find that the best things you might have thought about will be picked up from under your own nose by more venturesome spirits; but, above all, because this is the only manly thing to do.

I am lucky to have been born and to have grown up before the First World War, at a period at which the vigor and *élan* of international scholarship had not yet been swamped by forty years of catastrophes. I am particularly lucky that it has not been necessary for me to remain for any considerable period a cog in a modern scientific factory, doing what I was told, accepting the problems given me by my superiors, and holding my own brain only *in commendam* as a medieval vassal held his fiefs. If I had been born into this latter day feudal system of the intellect, it is my opinion that I would have amounted to little. From the bottom of my heart I pity the present generation of scientists, many of whom, whether they wish it or not, are doomed by the "spirit of the age" to be intellectual lackeys and clock punchers.

Have I gained or lost from my father's unconventional training? I do not know, for I have had only one life to live. My conjecture is that under a more conventional and milder regime I might have come through with less emotional trauma, but that I would not have developed the strong individuality of my scientific vein, which was due to early contact with a very powerful and very individualistic man. It was this struggle to maintain my individuality in the presence of a tremendously vigorous father which certainly gave the very specific form to my work which it later assumed.

While I might have achieved something under another training, one thing is clear: that without any training and guidance at all, my career would have been hampered and my productivity would have been distorted. It is very easy for a constitutionally vigorous mind to fritter its power away in trivialities. I put the highest value on my early contact with the standards of the intellect, and even though quite a different contact might have set me up as a scholar in another way, the absence of contact would have left me an ineffectual crank. I know of a number of cases where the relative paucity of scientific contacts, while not absolutely fatal, was still damaging and limiting.

A scientist must know what is being done in order that the very individuality of his own work may come to full fruition. He must live in a world where science is a career, where he has companions with whom he can talk and in contact with whom he may bring out his own vein.

It may well be true that ninety-five per cent of the really original scientific work is being done by less than five per cent of the professional scientists, but the greater part of it would not be done at all if the other ninety-five per cent were not there to help create a high level of scientific opinion. Even the self-trained scholar must pay tribute to that atmosphere of disinterested scholarship created by the universities, which furnishes the frame within which he may operate.

There is no doubt that the present age, particularly in America, is one in which more men and women are devoting themselves to a formally scientific career than ever before in history. This does not mean that the intellectual environment of science has received a proportionate increment. Many of today's American scientists are working in government laboratories, where secrecy is the order of the day, and they are protected by the deliberate subdivision of problems to the extent that no man can be fully aware of the bearing of his own work. These laboratories, as well as the great industrial laboratories are so aware of the importance of the scientist that he is forced to punch the time clock and to give an accounting of the last minute of his research. Vacations are cut down to a dead minimum, but consultations and reports and visits to other plants are encouraged without limit, so that the scientist, and the young scientist in particular, has not the leisure to ripen his own ideas.

Science is better paid than at any time in the past. The results of this pay have been to attract into science many of those for whom the pay is the first consideration, and who scorn to sacrifice immediate profit for the freedom of development of their own concepts. Moreover, this inner development, important and indispensable as it may be to the world of science in the future, generally does not have the tendency to put a single cent into the pockets of their employers.

Perhaps business has learned to take long risks, but they must be calculable risks, and no risk, by its very nature, is less calculable than the risk of profit from new ideas.

This is an age in which the profit motive is exalted, often, indeed, to the exclusion of all other motives. The value of ideas to the community is estimated in terms of dollars and cents, yet dollars and cents are fugitive currency compared with that of new ideas. A discovery which may take fifty years before it leads to new practice has only a minimal chance of redounding to the advantage of those who have paid for the work leading up to it, yet if these discoveries are not made, and we continue depending on those which already exist, we are selling out our future and the futures of our children and grandchildren.

Like a tradition of scholarship, a grove of sequoias may exist for thousands of years, and the present crop of wood represents the investment of sun and rain many centuries ago. The returns of this investment are here, but how much money and how many securities remain in the same hands, even for one century? Thus, if we are to measure the long-time life of a sequoia grove in terms of the short-time value of money, we cannot afford to treat it as an agricultural enterprise. In a profit-bound world, we must exploit it as a mine and leave a wasteland behind us for the future.

There are scientific ideas which we can trace clearly to the time of Leibniz, a quarter of a millennium ago, which are just beginning to find their applications in industry. Can a business firm or a government department, moved primarily by the immediate needs for new weapons, compass this period of time in its backward glance?

The great grove of science must be left to long-time institutions capable of formulating and maintaining long-time values. In the past, the Church was one of these institutions, and, even though it has somewhat fallen from its high estate, it has given birth to the universities and other intellectual institutions, such as academies, which themselves have a continuous life lasting over centuries.

These long-time institutions cannot and do not ask for an immediate translation of their hopes and ideals into the small change of the present day. They exist on faith, the faith that the development of knowledge is a good thing and must ultimately conspire in the good of all men.

The problem of planning for a long future is not unknown to business, particularly to that most sophisticated one the insurance business. The art of the actuary is to estimate risks. But the insurance business in general is not concerned merely with protection against destructive risks. The same companies that sell insurance also sell annuities. In a similar way, any long-time planning for the future must involve the discussion of rare and incalculable favorable circumstances as well as rare and incalculable catastrophes. One of the rare and incalculable benefits for which we must provide if the race is going to survive is the sudden emergence on the scene of great and original intellects.

A policy which integrates the gifts of the intellect into a long-time policy must transcend the lifetime of short-time institutions such as everyday business, and must be transferred to more stable institutions, like the foundations and the universities, which at least contemplate such a continued existence.

I am not alone in saying these things, but I am swimming counter to the major currents of the times. It is popular to believe that the age of the individual and, above all, of the free individual, is past in science. There are many administrators of science and a large component of the general population who believe that mass attacks can do anything, and even that ideas are obsolete.

Behind this drive to the mass attack there are a number of strong psychological motives. Neither the public nor the big administrator has too good an understanding of the inner continuity of science, but they both have seen its world-shaking consequences, and they are afraid of it. Both of them wish to decerebrate the scientist, even as the Byzantine State emasculated its civil servants. Moreover, the great administrator who is not sure of his own intellectual level can aggrandize himself only by cutting his scientific employees down to size.

The limiting case of the great scientific institution, by which we may test the soundness of the principles on which it acts, is the writing shop of the monkeys and the typewriters which, in the course of the ages, will almost certainly succeed in making every possible combination of the letters of the alphabet and the words in the dictionary. What is the real value of the work of the monkeys and the typewriters? Sooner or later, they will have written all the works of Shakespeare. Are we then to credit this mass attack with creating the works of Shakespeare? By no means, for before writing the works of Shakespeare, they will almost certainly have created just about all the nonsense and balderdash conceivable.

It is only after the non-Shakespearean has been thrown away, or at least an overwhelming part of it, that Shakespeare will stand out in any significant sense, whether theoretical or practical. To say that the monkeys' work will contain the works of Shakespeare has no other sense than to say that a block of marble will contain a statue by Michelangelo. After all, what Michelangelo does is purely critical, namely, to remove from his statue the unnecessary marble that hides it. Thus, at the level of the highest creation, this highest creation is nothing but the highest criticism.

Of course, the large laboratory can make out a limited case for itself. However, it is perfectly possible for the mass attack by workers of all levels, from the highest to the lowest, to go beyond the point of optimum performance, and to lose many really good results it might obtain in the unreadable ruck of fifth-rate reports. This is a real observable defect of large-scale science at the present time. If a new Einstein theory were to come into being as a government report in one of our super-laboratories, there would be a really great chance that nobody would have the patience to go through the mass published under the same auspices and discover it.

The great laboratory may do many important things, at its best, but at its worst it is a morass which engulfs the abilities of the leaders as much as those of the followers.

I have not found the great laboratory the medium in which I can develop my work with the freedom which I need to express its particular message. This may well be a limitation of my own nature, but my experience with young men has shown me that it is a limitation shared by many of those who have much to say. I hope and pray that the value of this important stratum of scientific workers will not be thrown away on the basis of short-sighted considerations of facility of administration and the trend of the times. Certainly I owe my own continued ability to do useful work to the cordial spirit of the administrators of M.I.T. and to their habit of protecting me from unwarranted claims on my time and from those who may have a narrow conception of my function.

After thirty-six years of functioning in the free atmosphere of M.I.T., and at the age of sixty, I do not find myself at the end of my scientific interests nor, I hope, of my achievements. My collaborative work on brain waves seems to me about to blossom out into a considerable science. Similarly, my joint studies with Armand Siegel on the Brownian motion and on time series is leading me to a reconsideration of the relative parts played in this world by cause and by chance. How many years may be granted me, if not to carry out this program of work myself, then at least to see that it is being carried out and to understand the share in it of my past ideas, I do not know; but even now I can feel reasonably sure that my scientific career, though it began early, is lasting late.

Index

Aberdeen Proving Gound, 29, 79
Acta Mathematica, 144, 318
Aerodynamics, 37
Affair of the Rites, 202-3
Agra, 354
Ahmadabad, 343-46
Aiken, Howard, 266, 297-99
Air Force, U. S., 312
Air India, 340-41
Ajanta, 349
Alexander, J. W., 27, 95
All-India Science Congress, 338, 341, 347
"Almost periodic functions," 93, 132-33
Alpha rhythm, 319
Alternating and direct current, 73-75

American Mathematical Society, 66, 85, 113, 115, 129, 174, 176, 225, 229-31
American Society of Mechanical Engineers, 309
American Colony at Göttingen, 116
American Telephone & Telegraph Co., 194
Amerika-Institut in Berlin, 114
Ampère, André, 322
Analogy computing machines, *see* Computing machines, high-speed
Analysis, mathematical, 28, 41, 175, 177, 190, 323
Anchor Books, 335
Andersen's *Fairy Tales,* 122
Annals of Mathematics, 145
Anti-aircraft artillery control data,

29, 143, 240-41, 249, 250-51, 254, 255, 260, 261, 325
Anti-Semitism, 212
Artin, 157
Ashkenazic and Sephardic culture, 151
Ataxia, 287-88
Atlantic Monthly, 296, 298
Atom-splitting, 154-55
Atomes, Les, 38-39
Atomic bomb, morality of, 295ff.; public opinion on, 301-7; theories of, 25, 143, 155, 293ff.
Atomism, 98, 101-2
Aurangabad, 349
Autocorrelator, 289-90
Automation, 275, 295, 308-13, 334; in India, 351, 354-56
Averages over curves, theories of, *see* Lebesgue integral
Aztecs, 284

Babbage, Charles, 136, 266
Ballistic computation, 227, 256
Banach, Stefan, 60
Banach-Wiener space, theory of, 60, 63-64, 93
Bangalore, 351
Barnett, Irving, 35, 36, 174
Bell Telephone Laboratories, 133, 134, 135, 178-79, 229, 249, 263, 333
Benares, 354
Berger, Hans, 319
Berlin Institute of Technology, 45
Bernard, Claude, 291
Beurling, 317
Bhabha, 352
Bierwirth, Professor, 47

Bigelow, Julian, 242-44, 248, 253, 254
Binary scale, *see* Numeration
Birkhoff, G. D., 27-28, 29, 39, 40, 41, 85, 112, 114, 116, 128, 142, 217, 281
Bisonette, Mr. and Mrs., 161
Bit, the, 269
Blake, John, 339
Blake, Peggy Wiener, 128, 144, 147-48, 149-50, 159, 183-84, 187-88, 193, 196, 207, 223, 333, 335, 350, 351
Blaschke, Wilhelm, 157
Bôcher prize, 177, 180, 208
Bodet, Torres, 280
Bohr, Harald, 93, 119, 121, 122, 132-33, 317
Bohr, Niels, 104-5, 121, 122, 304
Boltzmann, Ludwig, 324
Bolyai, John, 52
Bombay, 351-52
Bombay Rotary Club, 353
Born, Max, 107-9, 111
Bose, Professor, 353
Boston University, 109, 333
Bouligand, G., 92-93, 94-95, 318, 334
Bourbaki, 316, 318
Brain waves and statistical theory, 288-91, 319-20, 365
British Colony at Göttingen, 116
Brown University, 130, 144
Brownian motion, 37-40, 41, 64, 71, 78-79, 93-94, 109, 179, 288, 365
Bureau of Standards, 139, 297
Bush differential analyzer, 136-37, 242
Bush, Vannevar, 111-12, 133,

136-37, 138, 139, 141-42, 179, 190, 231, 232, 235, 239, 242, 293, 322

Cairo, 204
Calculus of variations, 36, 245
Calcutta, 353-54
Calcutta, University of, 353
Caldwell, Professor, 242
California, University of, 297, 298
Calvin, John, 283
Cambridge Philosophical Society (Eng.), 159, 160
Cambridge University, 21, 22, 25, 48-49, 56-57, 67, 115, 146, 148, 149-51, 152-54, 155, 173, 180, 228, 315; see also Trinity College
Cambridge University Press, 156
Cameron, Robert, 179-80
Cannon, Walter, 171, 221, 253, 291
Cardinal of India, 341
Carleman, 317-18
Carnegie Institute of Washington, 111
Cartwright, Miss, 152, 153
Caton, Richard, 319
Cauer, Mr. and Mrs. Richard, 142
Celestial mechanics, 217
Chandrasekharan, 352
Chao, Y. R., 201-2
Charcot, Jean, 18, 213
Chicago, University of, 32, 66, 217, 316
Child, Francis, 46
China, 184-203, 339
Chinese-Japanese War, 194-95, 218, 273
Christie, Agatha, 349
Clonus, 277, 278-79, 286

Cockcroft, 154
Collège de France, 333
College of St. Francis Xavier (Bombay), 353
Colloquium Series, 177-78
Columbia University, 26, 27, 50, 217, 293
Communication theory, 260, 262-63, 264, 265, 315, 321-27; in animals, 326; in England, 315; see also Cybernetics
Communications engineering, 72-79, 255, 264-65, 274, 325; in England, 151; see also Electrical engineering, theoretical
Communism, 218-21, 284; Chinese, 220; English, 206-7; see also U. S. S. R.
Comptes Rendus, see French Academy of Sciences
Compton, Karl Taylor, 140-41, 209, 218
Computing laboratories, 248-51
Computing machines, high-speed, 136-39, 178-79, 227, 232-35, 238, 242, 243-44, 266-68, 269, 290-91, 295, 297, 308, 315, 325, 334; analogy and digital, 137, 138, 190, 233ff., 290; see also Bush differential analyzer
Confucian tradition, 197-98
Congress, U. S., 304
Consolidated Edison Co., 74
Constructionalism, 51-54
Continuistic theory, see Atomism
Continuous spectrum theory, see Harmonic analysis
Control theory, see Communication theory; Cybernetics
Copenhagen, University of, 121

Corliss, Janet, 126
Corliss, Mr. and Mrs. Louis, 125-26
Cornell University, 21
Corposant, 81
Courant, Richard, 96, 108, 113, 114, 116, 117, 121, 211
Cramer, 64, 262
Cybernetics, 269, 274-75, 308, 327-28, 331, 335; see also Cybernetics
Cybernetics, 315-17, 321-22, 325, 328-29; publication of, 331-32
Czechoslovak trials, 207

Dabo, 55, 65
Dagsburg, see Dabo
Daily Worker, 206
Daniell, P. J., 36, 91
Dartmouth College, 229, 231
Darwin, Charles, 18
De La Vallée Poussin, 120
Delhi, University of, 354
Destructive instincts, human, 300
Determinism, 34, 104, 107; see also Newton and Newtonian physics
Dickson, Leonard Eugene, 66
Dielectric strength, 82
Differential equations and geometry, 124-25, 137, 142-43, 153; and atomic bomb, 294
Digital computers, see Computing machines, high-speed
Dirac, Paul, 109
Doklady, 261
Doubleday & Company, 335
Douglas, Jesse, 208, 209
Drake, Sir Francis, 154
Drew University, 333
Dreyfus, Mrs., 199
Dufèze, M., 335

Ecole Polytechnique, 64
Ecole Supérieure de Télécommunications, 334
Econometrics, 260
Eddington, Arthur Stanley, 100n.
Editions des Deux Rives, 335
Einstein, Albert, 25, 38, 54, 104, 175, 211, 281, 302, 304, 364
Eisenhart, 66
Electrical circuit theory, 139, 142, 168, 189-90, 192, 236
Electrical computing machines, 111-12; see also Computing machines, high-speed
Electrical engineering: in industry, 135; problems in World War II, 227; theoretical, 72-75, 168-69; see also Communications engineering; Power engineering
Electroencephalographs, 288, 289-90, 319, 320
Electromotive forces, 80-82
Electron theory, 25
El Greco, 284
Ellora, 349
Emotional symbols of mathematics, 86; see also Mathematical aesthetics
Employment and unemployment, 295-96, 311-12; see also Labor and management
Encyclopedia Americana, 29
Epilepsy, 289, 319
Erdös, 228
Ergodic theorem, 142, 217
Erro, Señor, 280
Espionage, see Security problems, U. S.
Ether theory, 103-4
Euclidian geometry, 51, 54

European colonialism, 303
Everest, Mt., 346
Existentialism, 324-25, 328, 334
Ex-Prodigy, 17, 86, 336

Factory workers, *see* Labor and management
Fascism, 163, 202, 220, 226, 284
Federal Institute of Technology (Zurich), 163
Feedback mechanisms, 190, 254, 264, 268-69, 290; negative, 252-53, 254, 291; in nervous and neuromuscular systems, 277, 290, 320; in office dictation, 331; *see also* Cybernetics; Statistical theory and mechanics
Fermi, Enrico, 211, 304
Finch, G. I., 346
Fire-control apparatus, 240-41, 264; *see also* Anti-aircraft artillery control data
Flip-flop circuits, 236-37, 268
Fluctuating currents and voltages, theory of, 73-74
Fluid flow, theory of, 94
Fort Monroe, Va., 249
Fourier Series and integral, 42, 76-77, 133, 139, 156, 168, 178, 180
Fourier Transforms in the Complex Domain, 178
Franco Spain, 206
Frank, Philipp, 157
Franklin, Philip, 79
Fréchet, Maurice, 41, 50, 54-55, 56, 58, 59-60, 64, 65, 334; theory of limits and differentials, 59-60
Frege, G., 54

French Academy of Sciences, 76; *Comptes Rendus* of, 92, 261
Freud and Freudians, 213-15, 216
Freymann, 315-17, 319, 321, 329, 331, 335, 340
Friends of China, 223, 350
Fuchs, Klaus, 108, 302
Fujiwara, Professor, 203
Fulbright grants, 333
Functions, theory of, 190-91

Gadgeteers, 305
Gándara, Nápoles, 279
Gandhi, 346
Gâteaux, M., 36, 64
Gauss, Karl, 52
General Electric Co., 29, 74-75
Generalized integration, theory of, 174
Geneva Canton Hospital, 336
Geodetic Survey of Greenland, 122
George, Clare, 125, 126
German Association for the Advancement of Science, 119
German University at Prague, 157, 158
Ghosts (Ibsen), 113
Gibbs, Josiah Willard, 23, 34-35, 38, 39, 71, 104, 142, 217, 255-56, 257, 260, 323, 324
Gilbreths, 311
Girton College, 152
Gödel, 157, 324
"Gold-Makers, The," 160
Göttingen, 21, 24, 30, 48, 95-96, 97, 105, 106, 107, 110, 113, 114-18, 119, 121, 124, 129, 162, 174
Greek geometry, 51
Greenhill, Sir George, 66

Group theory, 60
Guadalajara, 283, 284-85
Guggenheim Foundation and Fellows, 96, 113, 176, 281
Guided missiles, 296
Gulliver's Travels, 100n.-1n.

Hadamard, Jacques, 67-68, 120, 191, 193, 198-99, 200, 204, 334
Hahn, Otto, 157
Hahn, Paul, 291
Haldane, J. B. S., 160, 161, 162, 205, 206-7, 208, 314-15
Hardy, G. H., 21-22, 23, 24, 26, 41, 57, 113, 115, 123, 152-53, 156, 180, 212
Harmonic analysis, 76-79, 93, 97, 105-6, 115-16, 122, 139, 144-45, 190, 191, 217, 264, 288, 289, 314, 318, 323; optical computer for, 112
Haruna Maru, S. S., 202
Harvard Library, 352
Harvard Mathematics Club, 30
Harvard Medical School, 171-72, 253, 286, 288
Harvard Observatory, 206
Harvard University, 18, 21, 26, 27, 30, 31, 45-46, 47, 48-49, 57, 58, 66, 71-72, 79, 84, 130, 132, 141, 142, 157, 166, 212, 222, 266, 281, 287, 297, 298
Ha-Ta-Ma Hsien-Sheng, 193
Heat-death, 324, 328
Heaviside, Oliver, 78, 265; calculus of, 78, 93
Heilbronn, 135
Heisenberg, Werner, 107-8, 109, 244
Herman et Cie., 315, 316, 340

Hermite, Charles, 199
Hilbert, David, 24-25, 96, 97, 108, 109, 175, 192
Hiong, Professor and Mrs., 186, 188-89
Hiroshima, 299, 306
Hitler, *see* Nazis and Nazism
Hölder, 217
Hollerith machine, 237
Homeostatic processes, 291
Hopf, Eberhard, 142-43, 208, 209-11
Hopf-Wiener equations, 143, 245, 246
Houghton Mifflin Co., 335
Hu, Lottie, 223
Human Use of Human Beings, The, 335
Humboldt Library, 18
Huntington, Edward Vermilye, 52, 53
Hyderabad, 338, 348-49, 350
Hydrogen atom theory, 104-5

Ikehara, Shikao, 133, 135, 180, 183, 184
Illinois Institute of Technology, 157, 175
Imaginary numbers, theory of, 75
Imperial Hotel (Tokyo), 183-84
Indeterminism, 104, 107; *see also* Determinism; Probability and statistics
India, 337, 338-56
Indian Academy of Sciences, 344
Indian Atomic Energy Commission and Institute, 341
Indian Congress Party, 343
Indian Institute of Statistics, 353
Indian Mutiny, 342

Induction motor, see Alternating and direct current
Industrial Revolution, 310
Information theory, 178-79, 325
Ingham, A. E., 115, 116, 228, 229
Institute for Advanced Study, 175, 177
Instituto Nacional de Cardiología, 277, 282-83, 286, 321, 334
Interferometer, 289-90
International Business Machines Corp., 237, 238, 242
International Mathematical Congress: 1920 (Strasbourg), 49-50, 55, 65-70; 1924 and 1928, 145; 1932 (Zurich), 162-64; 1936 (Oslo), 200, 201, 203, 204; Cambridge, Mass., 335
International scientific cooperation, 304
Iron lung, 288
Italy, 118, 119

Jackson, Dugald C., 72, 73, 77
Janet, Pierre, 18, 213
Japan, 183-184; in World War II, 303-5, 339; see also Chinese-Japanese War; Hiroshima
Japanese chess, 203
Jessen, 317
Johns Hopkins University, 283
Johnson, Martin and Osa, 69
Jordan, Camille, 66-67
Juárez, Benito, 279
Juniata College, 111, 113

Kakutani, 184, 228
Kansas, University of, 66
Kantian theory of space, 54
Kapitza, P. L., 155

Kellog, O. D., 79-80, 82-84, 85
Kerensky, Alexander, 48
Khintchine, 145
Kiel, University of, 119, 120
Kierkegaard, Sören, 324, 328
Kinetic theory of gases, 102-3, 217
Kingsley's Natural History, 18
Klein, Felix, 30, 96-97, 113
Kline, J. R., 115, 116, 118, 128, 175
Knowledge, theory of, 327-28
Koebe, 159
Kolmogoroff, 145, 261-62, 325
Koopman, Bernard, 39, 217
Korean War, 203
Kosambi, Professor, 352
Kreuger, Ivar, 131
Krishnan, Professor, 354
Ku, Dean, 181, 201
Ku-Klux-Klanism, 220
Kuomintang, 219, 220

Labor and management, 275, 296, 308-13; in India, 354-56
Laënnec, René, 283
Lagrange, Joseph, 217
Landau, Edmund, 24, 135
Language and writing, 326-27
Laplace, Pierre, 217
Lawrence, D. H., 154
Learning, 291
Lebesque integral, 22-23, 33, 36, 37, 38, 39, 77, 92, 142
Lee, Dr. Yuk Wing, 133-34, 135, 142, 180-81, 185-86, 187, 189-90, 193, 195, 199, 218, 273-75, 289
Lee, Mrs. Yuk Wing, 186, 187, 218
Leeds, University of, 115
Leeuwenhoek microscope, 98-99
Lefschetz, Solomon, 66

Leibniz, 36, 80, 100, 100n., 101-2, 328, 362
Leipzig, University of, 94, 159
Lemaître, Canon, 206
Leukemia, 291, 292
Levinson, Norman, 180, 211, 212, 232, 255
Lévy, Paul, 41, 64, 90, 262, 318
Lichtenstein, Leon, 94, 95, 159, 163, 174, 209, 216
Lichtenstein, Mrs. Leon, 95
Littlewood, J. E., 23-24, 115, 152-53, 167
Lobachevski, Nikolai, 52
London, University of, 315
Los Alamos, 305
Low temperature physics, 155
Lowell, Abbot Lawrence, 58, 222
Lysenko, 207

McLaurin, Richard, 139, 141
Macy Foundation, 285-86
Madras, 343, 350-51
Madrid, 335-36
Mahalanobis family, 353, 354
Maine, University of, 29
Malin, Henry, 180
Management, *see* Labor and management
Manchester, University of, 315
Mandelbrojt, Szolem, 191, 200, 204-5, 317, 318, 333
Manhattan Project, 294-95, 299
Marconi, Guglielmo, 72
Margain family, 279
Martin, W. T., 179-80
Masani, 352
Masaryk, Thomas, 47, 158-59
Massachusetts General Hospital, 289, 320

Massachusetts Institute of Technology, 17, 30-31, 61, 71, 90, 91, 110, 124, 126, 129, 130, 132, 139-41, 142, 156, 168, 171, 174, 178, 179, 207, 208, 209, 210-11, 226, 228, 231, 242, 257, 273, 275, 279, 286, 287, 289, 297, 315, 320, 329, 333, 353, 365; administration of Compton, 140-41
Mathematical aesthetics, 60-63, 201
Mathematical collaboration, 200-1
Mathematical Institute (Hamburg), 158
Maximilian, 279
Maxwell, Clerk, 102-3, 217, 324
Mayan race, 236
Measure, theories of, *see* Probability and statistics, theories of
Mei, President, 181
Mejía, Gen., 279
Memory, 269, 291; of society, 326
Menger, Karl, 157, 175
Message coding and decoding, 179, 239-40, 302
Meteorology, 259-60
Mexico, 276-92, 316, 321, 328-29, 337, 340
Mexico City Nutrition Laboratory, 281
Mexican Mathematical Society, 276, 283-84
Mexico, University of, 279
Michelangelo, 364
Michelson, Albert, 104, 289, 290
Military expenditures, 294-95, 304-5
Military philosophy, 299
Milyukov, Paul, 47
Minnesota, University of, 180

"Miracle of Purun Bhagat," 345
Missouri, University of, 45
Möbius's sheet, 26
Monte Carlo method, 238-39
Montemar, Gen., 279
Moore, C. L. E., 31
Morison, Dr. Robert, 286
Morley, Edward Williams, 104
Morris, William, 310
Morse, Professor, 177
Motion study, 311
Mountbatten, Lord and Lady, 343, 346
Muckenhoupt, Carl, 132-33
Munich, 225
Murnaghan, Professor, 283-84
Murray, Forrest, 66
Murray Hill Hotel (N.Y.C.), 113
Museum of the Ecole Centrale des Arts-et-Métiers, 57
Muscio, Dr. and Mrs. Bernard, 56, 64
Musical notation, 105-6
My Friend Mr. Leakey, 207
Mylapore, 354

Nancago, University of, 316
Nancy, University of, 314, 316, 317-19, 335
Nanking, 201-2
National Academy of Sciences, 176
National Chemical Laboratory (India), 346
National Physical Institute (India), 354, 355
National Physical Laboratories (England), 315
Nature, 129
Navy Department, U. S., 297
Nazis and Nazism, 114, 116, 157, 159, 174, 176, 186, 209, 212, 219, 220, 226, 228, 295, 301, 304
Nehru, Jawaharlal, 341, 343, 346, 356
Nerve spike, theory of, 287
Nervous system and communication, 267-69, 285-86, 287, 290-91, 320, 323-24; see also Feedback mechanisms
Neumann, John von, 39, 109, 175, 192, 211, 217, 243, 259
Neurophysiology, see Brain waves; Harvard Medical School; Nervous system
New York Times, 156
Newton and Newtonian physics, 28, 34, 36, 107, 255, 257-58, 267, 327
Ni family, 187
Nichols, Ernest, 139
Nobel Prize, 107
Noether, Emmy, 162-63, 175
Non-Euclidian geometry, 52
Nørlund, Professor and Mrs., 122
Notre Dame University, 157, 175
Nuclear research and theory, 155, 300, 301, 304; see also Atomic bomb
Number theory, 120, 121
Numeration, 230-31, 236

Olvera, 282
Onnes, Kammerlingh, 155
Operators, theory of, 241-42
Organization theory, 315, 323; see also Communication theory; Cybernetics
Orozco, José, 284
Osaka, University of, 183, 184

Osgood, F. W., 30-31
Ostrowski, 317
Oxford University, 21, 48-49, 57, 67, 113, 153, 154, 319

Paley, 152, 162, 163-64, 167-69, 170, 173, 177-78, 190, 191-92, 315
Palmieri, Father, 48
Parallel postulate, 51-52
Pasteur, Louis, 199
Patents and inventions: in England, 151-52; in U. S., 134, 195, 267
Pauli, Wolfgang, 109
Pearl Harbor, 272-73
Pearson's (pub.), 160
Peiping, 185-201
Peiping Union Medical College, 198
Peissner, Fritz, 63, 70, 77, 79
Pennsylvania, University of, 115, 175
Perrin, Jean Baptiste, 38, 39
Phillips, Henry Bayard, 33-34, 91, 132
Philosophy of science, 327
Physical mathematics, 33, 38; *see also* Gibbs, Josiah Willard
Pittsburgh, University of, 66
Plancherel, 77, 317
Planck, Max, 97, 104
Poincaré, Henri, 28, 41, 192
Pólya, 175, 182
Poona, 346
Postulationalism, 27, 51-55
Potential theory, 80-83, 90-91, 95; *see also* Lebesgue integral
Power engineering, 72, 73, 264-65;

see also Electrical engineering, theoretical
Prediction, theory of, 143, 241, 244-45, 246, 249, 250-51, 261, 262, 297, 308, 321, 325, 346, 352
Predictors, 293-94
Prime-number theory, 135-36
Princeton University, 31, 66, 83, 95, 140, 141, 157, 175, 189, 269, 297; *see also* Institute for Advanced Study
Principia Mathematica, 51, 52
Probability and statistics, theories of, 23, 33, 34-36, 142, 145, 257-60, 323; *see also* Lebesgue integral, Monte Carlo method
Proceedings of the London Mathematical Society, 36
Proceedings of the National Academy of Sciences, 71
Profit motive, 361
Pythagorus, 75-76, 105

Quantum mechanics and theory, 54, 97-98, 104-5, 107-9, 125, 346; background of, 98-104
Quantum theory of atom, 108-9
Quasi-analytic functions, 190, 192

Racial relations, international, 301, 303-5, 353
Radar, 227-28, 246-48, 251, 312
Radcliffe College, 32, 223, 333
Rademacher, 175
Radio broadcasting and radiotelephony, 72-73, 264; *see also* Communications engineering
Radioactive isotopes, *see* Uranium isotopes

Raisbeck, Barbara Wiener, 127, 128, 146, 149, 159, 183-84, 187-88, 193, 207, 223-24, 321, 333, 350
Raisbeck, Gordon, 333
Raman, Sir C. V., 344, 346, 351
Raman, Lady, 346
Ramos, García, 280-81
Relativity, theory of, 25, 54, 104, 109, 281
Remington Rand, Inc., 309
Renou, Father, 202, 203
Reports of the Russian Academy of Sciences, 261
Reuther, Walter, 309
Rice Institute, 36
Richardson, R. G. D., 130, 144
Rickshaw boys, 192-93
Riemann, George, 120
River, The, 353
Rivera, Diego, 282-83
Rockefeller Foundation, 114, 176, 286
Rockefeller Institute, 249
Roman Catholic Church, 362
Rosenberg, Julius and Ethel, 302
Rosenblueth, Arturo, 170-72, 253-54, 269, 277-78, 281, 282, 283, 284, 285-86, 287, 288, 290, 292, 319, 321, 328-29, 337
Royal Society of England, 100n.
Royce, Josiah, 52
Russell, Bertrand, 21-22, 25, 26, 50-51, 52-53, 152, 324

Saccheri, 51
Sacco-Vanzetti case, 132
Salpêtrière, 18
Sarabhai, Vikram, 344, 346
Sarabhai, Mme. Vikram, 344

Sartre, Jean Paul, 334
Saslow, Samuel, 180
Scanning, 137, 232
Schmidt, Robert, 119, 120-21, 122
Schröder, E., 54
Schrödinger, Erwin, 109, 125
Schuster, Sir Arthur, 77
Schwartz, Mr. and Mrs. Laurent, 318, 335
Science Service, 333
Scientists, wartime mobilization of, 231-32
Second Jungle Book (Kipling), 345
Secrets, military, see Security problems, U. S.
Security problems, U. S., 301-3, 308
Servetus, 283
Servomechanisms, 252, 265, 269, 275; see also Cybernetics, Feedback mechanisms
Shakespeare, 364
Shannon, Claude E., 178-79, 263, 264, 315, 325
Shannon-Wiener definition of quantity, 263
"Shot effect," electronic, 40, 264
Siegel, Armand, 109, 365
Sinusoids, 76-77
Smith College, 90
Smoluchowski, 38
Society for the Advancement of Management, 309
Solar energy, 354
Space, co-ordinate representation of, 59
Spanish Armada, 154
Spanish Jesuits in India, 353
Spanish Loyalists, 221
Spermatozoa, 99-100

Springer's mathematical series, 255
Stanford University, 175, 182
Star radiation, 142-43
Statistical theory and mechanics, 23, 34, 38, 104, 244-45, 255, 262-63; of communications engineering, 274, 325; of nerve impulses, 287; *see also* Brain waves; Probability and statistics, theories of; Quantum theory
Steinmetz, Charles P., 75
Stone, 228
Strand, the (pub.), 160
Strasbourg, 49, 55, 58, 65-68, 89, 122
Stratton, Wesley, 139
Struik, Dirk Jan, 124-25, 133, 279
Suez, 204
Sunnyasi, 344-45
Swift, Jonathan, 99, 100n.-1n.
Switching devices, 178, 263, 265, 268; human beings as, 296
Synapses, 268, 287
Syracuse University, 179
Szász, Mr. and Mrs., 174-75
Szegö, 175, 182
Szilard, Leo, 211, 304

Tagore, Rabindranath, 353
Taj Mahal, 354
Taj Mahal Hotel (Bombay), 341
Tamarkin, J. D., 129-30, 144, 145, 177
Tata Institute, 351, 352
Tauberian theorems, 115-16, 119-20, 208
Taxco, 329
Taylor, Frederick W., 311
Taylor, Sir Geoffrey, 36-37, 41
Taylor, James S., 66, 68
Technology Press (M.I.T.), 331

Telephone and telephone engineering, 73, 75, 178, 229, 247, 263-64, 265, 312
Television, 73, 137, 138, 247-48, 264, 265
Tesla, Nikola, 74
"Theory of Ignorance, The," 324
Thermodynamics, second law of, 324, 328
Time series, 288, 352, 365
Time study, *see* Motion study
Tokyo, University of, 184
Topology, 26-27, 40-41, 50
Touraine, La, S. S., 55-56, 65, 69
Tracking, target, 251-252
Trinity College (England), 150, 153, 154, 160; *see also* Cambridge University
Tsing Hua University, 181, 186, 188, 189, 191, 193, 201
Tufts College, 21, 333
Turbulence, theory of, 36-37
Turing, 315
Trajectories, computation of, 153
Tyler, H. W., 31

Unions, *see* Labor and management
United Auto Workers, 309
University College (London), 314
Uranium isotopes, 293-95
Urey, Harold, 293-94
U. S. S. R., 70, 155, 201, 226, 347-50; and U. S., 301-2, 304; *see also* Communism

Vallarta, Manuel Sandoval, 92, 111, 170, 171, 206, 273, 276, 279, 344
Vallarta, Maria Luisa, 206, 279
Van Kampen, E. R., 217-18

Veblen, Oswald, 27, 40, 112, 175, 228
Vectors, theory of, 59, 63, 64
Vedic ethics, 344-45
Verne, Jules, 270
Vibrating systems, theory of, 75-76, 105, 191
Vijayaraghavan, 350-51
Von Mises, 157, 211

Walsh, Joe, 66
Walter, Grey, 319, 320
Walton, 154
Wave filters, 247-48
Waves and vibrations, see Fourier Series and integral
Weaver, Warren, 249, 286
Wells, H. G., 205, 270
Westinghouse Co., 66, 74-75
Weyl, Hermann, 175, 177, 192
Whig party (England), 100n.
Whitehead, Alfred North, 27, 50-51, 52-53, 57
Whitehead, Jessie, 57, 150
Wiener, Barbara, see Raisbeck, Barbara Wiener
Wiener, Bertha, 32, 90, 94, 95, 118, 145
Wiener, Constance, 32, 90, 123
Wiener, Fritz, 90
Wiener, Leo, 18ff., 25, 28, 30, 32, 43, 47-48, 55, 79, 85, 89-90, 114, 117-18, 126, 127, 165-67, 181, 212, 222; influence on Norbert, 18-19, 20-21, 357-58; intellectual interests, 46-47, reputation at Harvard, 44-46; youth and education, 44-46
Wiener, Mrs. Leo, 20, 32, 117-18, 126, 165

Wiener, Margaret Engemann (Mrs. Norbert Wiener), 84-85, 86-87, 110-11, 113, 115, 116-19, 121, 122-24, 125-27, 144, 146, 149, 158, 159, 160, 165, 166, 184, 193, 199-200, 201, 203, 204, 205, 207, 208, 212, 222, 223, 225, 292, 321, 335-36
Wiener, Norbert, 17ff.; aids displaced European refugee scholars, 174-76; 211, 216, 228; books read as child, 18; Brown lectures, 141; Cambridge lectures on Fourier integral, 156; childhood and adolescence, 17-19; college and graduate degrees, 21
comments on:
availability for consulting, 332-33; bringing up children, 223-24; college professors, isolation from world affairs, 132; competitiveness among mathematicians, 83-84, 87-88; faculty salaries, 130-32, 209; French, German and English university life, 67; German intellectual tradition, 46; intellectual and moral responsibilities of modern scientists, 306-08, 358-65; lecture tours, 332; mathematical aesthetics, 60-63; mathematicians vs. engineers, 266-67; medical theory of today, 291-92; need for East-West understanding, 339-40; present-day scientists and engineers, 270-72; public attitude towards scientists, 307-8; science fiction, 160, 270; secretaries and dictation, 331; student bull sessions, 271;

summer vacations for teachers, 127; warfare, present and past, 300-1 consciousness of Jewishness, 20; dispute with Bohr on "almost periodic functions," 93; dispute with Kellog on potential theory, 80-84; experiments in topology, 26-27; eye operations, 329-30; first pleasures in math with Hardy, 22; first problems in probability, 35-36; Harvard lectures on Whitehead, 27; inventions of, 133-35, 151; marriage and honeymoon, 113-20; marital adjustments, 84-87, 116-19; M.I.T. appointment and promotion, 31, 141; motivation for studying math, 86; as novelist, 204; present work habits, 330-31; psychoanalysis of, 213-16; reaction to Paris life, 57

research activities:
with Bouligand on potential theory, 92; with Fréchet, 60, 63-64; with Jackson at M.I.T., 72ff.; with Philips on potential theory, 90-91; on prime-numbery theory and Tauberian theorems, 120-21, 122; on quasi-analytic function, 192; with Rosenblueth on physiology and cybernetics, 286-89; in World War I, 25, 29-30; in World War II, 225-75; *see also* Anti-aircraft artillery control data; Automation; Brain waves; Brownian motion; Clonus; Computing machines, high-speed; Cybernetics; Feed-

back mechanisms; Harmonic analysis; Harvard Medical School; Lebesgue integral; Massachusetts General Hospital; Nervous system; Statistical theory and mechanics

travels:
see Cambridge; Copenhagen; China; Göttingen; India; International Mathematical Congress; Italy; Japan; Madrid; Mexico; Strasbourg
Wiener, Peggy, *see* Blake, Peggy Wiener
Wiesner, Dr. Jerome, 287
Wildes, K. S., 218
Wiley, John, and Sons, 331
Wilson, E. B., 71-72
Wintner, Aurel, 216-17
Woolwich, 66
Worcester Foundation for Experimental Biology, 333
World War I, 25-26, 47, 48, 68, 69-70, 114, 149, 153, 173, 225, 227, 240, 312, 339
World War II, 153, 157, 220, 225ff., 293, 299ff., 312, 339
Wright, Frank Lloyd, 183

Yale University, 34, 71
Yatsevich, Michael, 48
Yenching University, 187-88
Yokahama, 183-84
Yoshida, 184
Young, Mr. and Miss, 152

Zaremba, 83, 91-92
Zionist movement, 176
Zurich Federal Institute of Technology, 317